Hubert Brandstätter

Complement Factor H In Intestinal Ischemia Reperfusion Injury

Hubert Brandstätter

Complement Factor H In Intestinal Ischemia Reperfusion Injury

Mechanism Of Action

Südwestdeutscher Verlag für Hochschulschriften

Impressum / Imprint
Bibliografische Information der Deutschen Nationalbibliothek: Die Deutsche Nationalbibliothek verzeichnet diese Publikation in der Deutschen Nationalbibliografie; detaillierte bibliografische Daten sind im Internet über http://dnb.d-nb.de abrufbar.
Alle in diesem Buch genannten Marken und Produktnamen unterliegen warenzeichen-, marken- oder patentrechtlichem Schutz bzw. sind Warenzeichen oder eingetragene Warenzeichen der jeweiligen Inhaber. Die Wiedergabe von Marken, Produktnamen, Gebrauchsnamen, Handelsnamen, Warenbezeichnungen u.s.w. in diesem Werk berechtigt auch ohne besondere Kennzeichnung nicht zu der Annahme, dass solche Namen im Sinne der Warenzeichen- und Markenschutzgesetzgebung als frei zu betrachten wären und daher von jedermann benutzt werden dürften.

Bibliographic information published by the Deutsche Nationalbibliothek: The Deutsche Nationalbibliothek lists this publication in the Deutsche Nationalbibliografie; detailed bibliographic data are available in the Internet at http://dnb.d-nb.de.
Any brand names and product names mentioned in this book are subject to trademark, brand or patent protection and are trademarks or registered trademarks of their respective holders. The use of brand names, product names, common names, trade names, product descriptions etc. even without a particular marking in this works is in no way to be construed to mean that such names may be regarded as unrestricted in respect of trademark and brand protection legislation and could thus be used by anyone.

Coverbild / Cover image: www.ingimage.com

Verlag / Publisher:
Südwestdeutscher Verlag für Hochschulschriften
ist ein Imprint der / is a trademark of
OmniScriptum GmbH & Co. KG
Heinrich-Böcking-Str. 6-8, 66121 Saarbrücken, Deutschland / Germany
Email: info@svh-verlag.de

Herstellung: siehe letzte Seite /
Printed at: see last page
ISBN: 978-3-8381-3852-7

Zugl. / Approved by: Wien, Medizinische Universität, Dissertation, 2013

Copyright © 2014 OmniScriptum GmbH & Co. KG
Alle Rechte vorbehalten. / All rights reserved. Saarbrücken 2014

TABLE OF CONTENT

1 ABSTRACT 4
2 INTRODUCTION 6
   2.1 COMPLEMENT SYSTEM 6
   2.2 COMPLEMENT FACTOR H 8
   2.3 ISCHEMIA REPERFUSION INJURY AND COMPLEMENT SYSTEM 11
3 AIMS OF THE STUDY 13
4 MATERIALS 14
   4.1 PREPARATION PROCESS OF FACTOR H 14
   4.2 MATERIALS FOR THE RAT MODEL 14
5 METHODS 16
   5.1 PREPARATION PROCESS OF FACTOR H 16
   5.1.1 Biochemical and Functional Characterization of Intermediates and purified, final CFH 17
   5.2 RAT MODEL OF INTESTINAL IRI 17
   5.2.1 Ethics Declaration 17
   5.2.2 Animals 17
   5.2.3 Surgical procedure 19
   5.2.4 Blood sampling 22
   5.2.5 Tissue dissection 23
   5.2.6 Administration of purified human CFH 24
   5.2.7 Histology and Light Microscopy 25
   5.2.8 Pathology and Villi Length Analysis 26
   5.2.9 Immunohistochemistry 27
   5.2.10 Immunofluorescence Microscopy 29
   5.2.11 Rat blood biochemistry 30
   5.2.11.1 Factor H activity – Hemolysis Assay 30
   5.2.11.2 Factor H recovery in rat plasma – ELISA 31
   5.2.11.3 Complement C3 activation status – Western Blot 31
   5.2.11.4 Complement Hemolytic Activity in plasma – CH50 Assay 32
   5.2.11.5 Complement C3 concentration of rat plasma – ELISA 33
   5.2.11.6 Total protein of rat plasma – Bradford Assay 33
   5.2.11.7 Hematology 34
   5.2.12 CFH expression in Caco-2 cell lysates – Western Blot 34
   5.2.13 Statistical Evaluation 35
   5.3 RAT MODEL OF RENAL IRI 35
6 RESULTS 37
   6.1 NATIVE, FUNCTIONAL CFH WAS PURIFIED FROM HUMAN PLASMA FRACTIONS 37

6.2 A RAT MODEL OF INTESTINAL IRI INDUCES LOCAL MUCOSAL TISSUE INJURY AND COMPLEMENT DEPOSITION.. 38
6.2.1 *No systemic complement activation induced in the rat model.* 44
6.2.2 *Detection of C3 levels of rat plasma* 44
6.2.3 *Detection of Complement Hemolytic Activity (CH50) of rat plasma* 45
6.2.4 *Determination of total protein content of rat plasma samples* . 45
6.3 INTESTINAL IRI INDUCED LOCAL COMPLEMENT DEPOSITION IN THE ISCHEMIC GUT 48
6.4 ADMINISTRATION OF HUMAN CFH DOES NOT SIGNIFICANTLY PROTECT RATS FROM INTESTINAL IRI 50
6.4.1 *Inflammatory cells did not significantly contribute to intestinal IRI* 52
6.4.2 *Administered human CFH persists in rat circulation* 53
6.5 HUMAN CFH ADMINISTERED INTO RATS SUBJECTED TO INTESTINAL IRI DID NOT LOCALIZE TO SITES OF INJURY 54
6.5.1 *Detection of endogenous CFH on rat tissues* 55
6.6 HUMAN CFH DID NOT TARGET TO INTESTINAL SITES OF INJURY IN IRI . 60
6.7 HUMAN CFH DID NOT TARGET RENAL SITES OF INJURY IN IRI 62
6.8 CACO-2 CELL MODEL OF HUMAN INTESTINAL EPITHELIAL CELLS EXPRESSES CFH 64
6.9 HUMAN CFH BOUND DYING INTESTINAL EPITHELIAL CELLS IN INTESTINAL IRI 66
6.10 INJECTED HUMAN CFH INTO RATS SUBJECTED TO INTESTINAL IRI MAINTAINED FUNCTIONALITY IN RAT CIRCULATION *IN VIVO* 68
6.11 HUMAN CFH ADMINISTRATION INTO RATS SUBJECTED TO INTESTINAL IRI PREVENTED LOCAL COMPLEMENT DEPOSITION AT ISCHEMIC INTESTINE...... 70

7 DISCUSSION 72
8 FUTURE DIRECTIONS 82
9 REFERENCES 87
10 LIST OF ABBREVIATIONS 112
11 PUBLICATIONS 114
12 LIST OF TABLES 115
13 LIST OF FIGURES 116

1 ABSTRACT

Complement factor H (CFH) acts as major regulator of the alternative pathway of complement and mutations and genetic polymorphisms in the CFH gene predispose to various human diseases. This study aimed to purify CFH from human plasma and to test it as therapeutic complement inhibitor in a preclinical model of a human disease, in which complement activation has been implicated. The involvement of complement-mediated inflammation and tissue injury has been extensively demonstrated in animal models of intestinal ischemia reperfusion injury (IRI).

First, a scalable purification process of CFH from human plasma fractions of industrial plasma fractionation was developed to supply a pathogen safe and functional CFH concentrate that was biochemically and functionally evaluated. The novel preparation process achieved significant depletion of truncated and dysfunctional CFH species and purification of largely native CFH.

Second, intestinal ischemia reperfusion injury in rat ileum was induced by complete and warm ischemia for 30 min followed by 60 min of reperfusion. Neither evidence for systemic complement activation nor tissue damage in peripheral tissues nor other intestinal segments than ileum in this rat model could be found. However, significant local complement deposition and local pathologic injury of the ileal mucosa was observed, providing a rationale to administer human CFH as therapeutic complement inhibitor. Therefore, rats were injected with human CFH prior and subsequently subjected to IRI to investigate the hypothesis that intravenous application of human CFH would protect from clinical manifestations of intestinal IRI. Two doses of human CFH were applied in order to investigate a potential dose response on mucosal protection.

Administration of human CFH completely diminished local complement deposition in the ischemic intestine. However, it did not significantly prevent local mucosal injury. Thus, the elicited local complement deposition was not responsible for the mucosal injury in this model. Based on this it was concluded that injected human CFH did not efficiently target the intestinal epithelium as major site of injury in this rat model. Likewise, injected human CFH was not recruited to renal sites of injury in a rat model of renal IRI. In contrast, human CFH largely maintained functionality and integrity after 90 min in rat circulation. Injected human CFH was able to regulate rat complement in plasma, complement deposition in the rat intestine, and bound to dying intestinal epithelial cells, thereby demonstrating functionality *in vivo*.

Thus, CFH therapy appears to be more suitable in conditions of uncontrolled complement activation in the circulation, which are associated with fluid phase regulation, e.g. paroxysmal nocturnal hemoglobinuria, atypical hemolytic uremic syndrome or CFH deficiencies.

2 INTRODUCTION

2.1 Complement System

Complement is an ancient and strongly conserved host defense component of innate immunity that acts as a rapid immune surveillance system to eliminate altered self and dangerous non-self molecular structures (Ricklin et al, 2010). The complement system comprises a network of over 40 proteins present in plasma or on host cell surfaces that activate each other by proteolytic cleavage similar to the coagulation cascade (Amara et al, 2010; Huber-Lang et al, 2006; Oikonomopoulou et al, 2012). Besides the elimination of infectious microbes, cellular debris, complement also orchestrates and "complements" adaptive immune and inflammatory responses (Carroll, 2004a).

Complement can generally be activated by various pathways that merge at its core component, complement protein 3 (C3) (Fig. 1). Activation via each of the pathways enables the formation of C3 convertase complexes that cleave C3 into its opsonizing part C3b and its proinflammatory and chemotactic component termed C3a. Properdin is regarded as the only positive complement regulator stabilizing C3 convertases. If complement activation progresses until the formation of the terminal pathway, the membrane attack complex is assembled leading to pore formation and lysis of the opsonized structure (Muller-Eberhard, 1986).

The classical pathway is often referred to as antibody-dependent, although it can also be activated by various other molecular patterns than immunoglobulin complexes. The molecule C1q activates the classical pathway by recognizing antigens complexed with immunoglobulin complexes or endogenous pattern recognition molecules like pentraxins, or by recognizing structures on microbial and apoptotic cells (Nauta et al, 2002). Genetic deficiencies of any classical pathway component were found to be associated with autoimmune diseases such as systemic lupus erythematosus (Carroll, 2004b), likely a result of impaired clearance of apoptotic particles or immune complexes, and the impairment of adequate humoral responses (Mayilyan, 2012).

In the lectin pathway, mannose-binding lectin (MBL) and ficolins function as pattern recognition molecules that predominantly recognize carbohydrate patterns on microbes. Deficiencies in MBL are among the most common causes for human immune-deficiencies rendering individuals prone to microbial infections and autoimmune disorders (Mayilyan, 2012).

The alternative complement pathway depends on spontaneous auto-activation due to continuous C3 hydrolysis ("C3 tick-over") allowing the formation of C3 convertases to promote complement amplification (Nilsson & Nilsson Ekdahl, 2012). The initiation mechanism of the alternative pathway involves properdin that recognizes pathogen- and damage-associated molecular patterns on non-self and apoptotic cells, respectively (Ricklin et al, 2010). The alternative pathway accounts for up to 90 % of total complement activation, even when initially triggered by the classical or lectin pathway (Harboe et al, 2006; Harboe & Mollnes, 2008; Harboe et al, 2004).

Complement is tightly regulated at each step of the cascade by soluble and membrane-bound proteins (Fig. 2). Complement dysregulation as a result of deficiencies or mutations of regulators leads to uncontrolled complement activation, tissue damage and inflammation especially at vulnerable sites of the glomerular basement membrane of the kidney and the Bruch's membrane of the retina (Zipfel & Skerka, 2009). Deficiencies in components like C3 give rise to increased susceptibility to various fungal, bacterial or viral infections (Tichaczek-Goska, 2012). The critical importance of complement for immunity is also supported the plethora of strategies pathogens have developed to evade complement attack like surface expression of binding proteins or protein mimicry of host structures to recruit complement regulators (Lambris et al, 2008; Schneider et al, 2009).

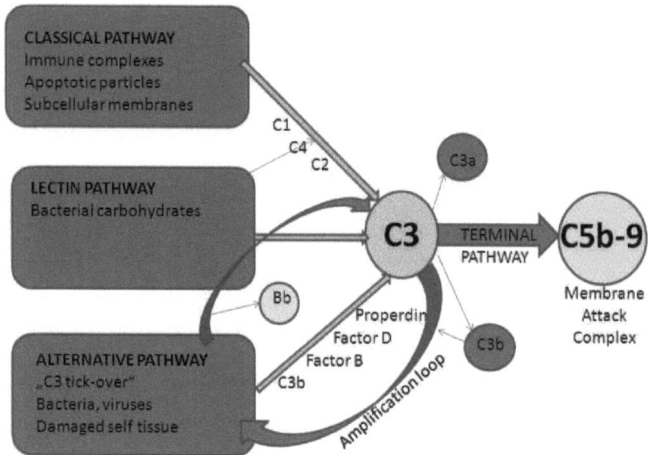

Figure 1 – The Complement System

2.2 Complement Factor H

Complement factor H (CFH) is an abundant serum glycoprotein that regulates the Alternative Pathway of Complement (AP) in fluid phase and on surfaces by various modes. First, it acts as essential cofactor for the serine protease complement factor I in cleaving C3b, the opsonin part of activated C3, into pro-inflammatory and chemotactic C3 fragments like inactivated C3b (iC3b) (Meri & Pangburn, 1990). CFH is not a cofactor for the cleavage of iC3b (Lambris et al, 1996; Ross et al, 1983). Second, CFH possesses decay acceleration activity for C3 convertases (Hourcade et al, 2002), which are enzymes in charge of complement amplification (Fig. 2).

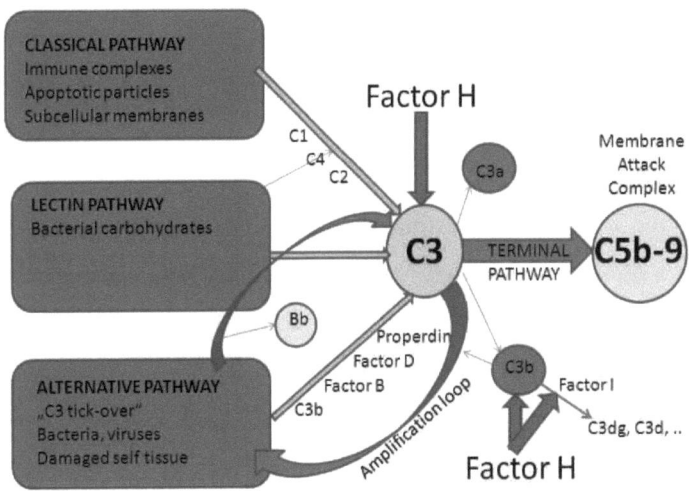

Figure 2 – CFH functions within the complement cascade

CFH is mainly produced in the liver, secreted into the circulation with a mean plasma concentration of 250 µg/ml (Hakobyan et al, 2008; Hakobyan et al, 2010). However, local production of CFH has been detected in various human tissues and cell types, presumably as means for local complement regulation in tissues naturally lacking endogenous membrane-bound complement inhibitors (Coffey et al, 2007) (Anderson et al, 2010).

The complete deficiency of CFH due to mutations, polymorphisms or anti-CFH antibodies is associated with atypical hemolytic uremic syndrome (aHUS) (Fremeaux-Bacchi et al, 2005) and membranoproliferative glomerulonephritis type II (MPGN II) (Smith et al, 2007) because it leads to uncontrolled AP activation in kidney and on blood cells and severe C3 deficiency (Pickering et al, 2006). These patients are treated by plasma exchange/infusion to replace the mutated, dysfunctional CFH molecules. Plasma therapy comprising plasma infusion or plasma exchange (Licht et al, 2007), however, bears major disadvantages including a high protein and volume load for the patients and additionally the theoretical risk of transmission of blood-borne pathogens. Therefore, patients with disease due to CFH deficiency could be treated more safely and efficiently with a CFH concentrate (Fakhouri et al, 2010a).

In addition to the replacement of mutated CFH and the substitution of complete CFH deficiency, another option for CFH therapy would be supraphysiological administration in pathologies that are associated with severe complement activation.

Aberrant activation of the alternative complement pathway has been found to be associated with several pathological conditions, namely age-related macular degeneration (AMD) (Anderson et al, 2010), membranoproliferative glomerulonephritis type II (Pickering et al, 2006), atypical hemolytic uremic syndrome (Davin et al, 2006),(Fremeaux-Bacchi et al, 2005) or renal ischemia-reperfusion injury (Thurman et al, 2003b). It has been suggested that specific AP inhibition could serve as potential therapy in diseases like rheumatoid arthritis or intestinal ischemia reperfusion injury involving aberrant complement activation due to the AP (Anderson et al, 2010; Banda et al, 2009; Huang et al, 2008b). Evidence has been provided that a tyrosine-to-histidine amino acid exchange at position 402 of CFH, with a prominent heterozygous prevalence of 30 % in healthy western populations, predisposes individuals to AMD (Edwards et al, 2005; Hageman et al, 2005a; Haines et al, 2005; Klein et al, 2005).

Having recognized the importance of CFH in human pathophysiology, multiple methods for the lab scale purification of CFH from mouse, rat and human plasma have been published (Alsenz et al, 1985; Ripoche et al, 1988; Sim & DiScipio, 1982b). Furthermore, methods for recombinant production of human CFH in moss and yeast (Buttner-Mainik et al, 2011; Schmidt et al, 2011) as well as methods for the industrial large scale production of human CFH from donor plasma have been published.

Claiming a potential future supply of human CFH as therapeutic or prophylactic agent for many individuals carrying CFH risk alleles, methods for scalable production of purified human CFH are of special interest in this respect.

To this point it was unclear whether highly purified human CFH can be isolated from plasma fractions of industrial plasma production. In particular, waste or side plasma fractions bear the unique advantage of an industrial scale source of human CFH without affecting validated plasma product processes. It was not known so far whether these materials – underlying protein precipitations due to variations of buffer, pH, temperature and ethanol addition - are suited for the purification of a fully functional human CFH protein.

2.3 Ischemia Reperfusion Injury and Complement System

A clinically important pathological setting implicating complement in the patho-mechanism is IRI, in which tissue ischemia followed by rescuing reperfusion initiates a paradoxically higher level of inflammation and tissue injury as occurs by ischemia alone (Banz & Rieben, 2011; Dorweiler et al, 2007). IRI occurs in many clinical-/surgical conditions, including organ transplantation (Asgari et al, 2010), myocardial infarction (Atkinson et al, 2010), cerebral stroke (Leinhase et al, 2006), major trauma or shock, limb ischemia, vascular surgery or cardiovascular bypass (Dorweiler et al, 2007).

Intestinal IRI causes impaired gut motility and mucosal wall injury as result of local inflammation due to complement activation, neutrophil infiltration and eicosanoid production (Hamar et al, 2003; Williams et al, 1999). Small bowel transplantation is inevitably associated with intestinal IRI (Mangus et al, 2009). As a major complication of intestinal IRI, bacterial translocation and sepsis (Atkinson et al, 2005b; Tsunooka, 2004) are mainly responsible for the high morbidity and mortality in trauma and surgical patients (Mallick et al, 2004).

Generally, complement becomes strongly activated after ischemia in several organs, including the heart (Busche & Stahl, 2010), the intestine (Fleming, 2003), the skeletal muscle (Pemberton et al, 1993), the brain (Harhausen et al, 2010) and the kidney (Thurman et al, 2003b), but the trigger/mechanism of complement activation seems to vary depending on the affected tissue. Binding of natural immunoglobulin M (IgM) antibodies to neo-epitopes on ischemic tissue is considered potential initiator of complement activation after IRI (Austen et al, 2004; Padilla et al, 2007; Zhang et al, 2004; Zhang et al, 2006). However, the mechanisms whereby the neo-epitopes become accessible to recognition remain to be largely understood (Kulik et al, 2009). Moreover, infiltration and activation of phagocytic cells in ischemic tissues undergoing reperfusion is regarded a key event of IRI, and phagocytes seem to be attracted by activated complement towards ischemic tissue (Grootjans et al, 2010; Hernandez et al, 1987; Simpson et al, 1993).

Complement inhibition by the monoclonal anti-C5 antibody eculizumab® proved successful in the prevention of antibody-mediated graft rejection in clinical cases (Lucas et al, 2011) and reinforces the concept of therapeutic complement inhibition in organ transplantation.

In mouse models of intestinal IRI, a set of recombinant mouse fusion proteins led to significant amelioration of IRI-induced complement activation and tissue injury of the intestine and lung (Atkinson et al, 2005b; Huang et al, 2008b; Rehrig et al, 2001). However, the complex mechanism by which the CFH-based complement inhibitor protected from intestinal injury in rodent IRI models is not yet entirely understood.

Detailed knowledge on how complement inhibition works in IRI will increase the potential applicability of complement therapeutics in these severe conditions.

3 AIMS OF THE STUDY

Ischemia Reperfusion Injury (IRI) is a clinically relevant pathological situation. Complement activation has been demonstrated to be a strong contributor to the patho-mechanism of reperfusion injury and complement interference has been shown to be protective.

Therapeutic complement intervention by CFH in a rat model of IRI may provide insights into the mechanisms of complement regulation by CFH *in vivo* and further help to understand the role of complement for the patho-mechanism of IRI *in vivo*.

The major hypothesis of this work was that administration of human CFH into rats subjected to intestinal IRI will provide protection. The major aims of this study were the identification of a novel and scalable purification process for CFH from human plasma and the evaluation of the therapeutic effect of human CFH in IRI.

4 MATERIALS

4.1 Preparation process of Factor H

Materials used for the preparation process of CFH and detection methods used to analyze CFH were described in Brandstaetter et al. (2012).

4.2 Materials for the rat model

Anesthesia was induced and maintained with Isofluran Baxter, Baxter Deutschland GmbH, Edisonstraße 4, 85716 Unterschleißheim, Germany. Additional analgesia during anesthesia was done by carprofen (Rimadyl®, Pfizer Corporation Austria, Vienna, ATCvet-code: QM1AE1). During the anesthesia electrocardiogram and body temperature were continuously monitored with a Small Animal Monitoring and Gating System. 1025 T (Small Animal Instruments, Inc., SAII, 65 Main Street, Stony Brook, NY 11790,www.i4sa.com).

The collateral vessels of ileum undergoing ischemia were ligated by sutures of Silk® braided black, 4-0 EP 1.5, SMI Hünningen 37, 4780 St. Vith, Belgium. The supplying mesenterial vessel of the ileal segment was ligated via a microvascular clamp (Mini Bulldog clamp CVD 35mm, lot 401392200, Aesculap AG & Co KG, Am Aesculap Platz, 78532 Tuttlingen, Germany). The closure of the abdominal wall to create warm ischemia was performed using Surgicryl® Polyglycolic acid, 4-0, EP 1.5, SMI Hünningen 37, 4780 St. Vith, Belgium. Paraffin for embedding of rat tissue used was Surgipath Paraplast (39601006, Leica Biosystems). Xylene and graded ethanol for histology in immunohistochemistry were purchased from Roth (www.carlroth.com) or VWR (https://at.vwr.com).

Whole blood after onset of (prevalue) and after 90 min of anesthesia of rats for hematology was collected in Vacuette®, 2 mL K_3EDTA, Greiner Bio-One GmbH, Haller Str. 32, 4550 Kremsmünster Austria. Rat plasma from healthy and completely untreated female Sprague Dawley rats obtained from lithium-heparinized whole blood immediately after anesthesia was used as "pre-immune rat plasma" for in vitro assays as described below.

In the rat model of renal IRI, urine excretion of rat kidney injury molecule-1 (KIM-1) in urine was detected by ELISA (RKM-100, R&D Systems, USA), levels of rat N-acetyl-β-D-glucosaminidase (NAG) in urine were likewise detected by ELISA (BlueGene Biotech, China). Blood urea nitrogen (BUN) was assayed using BUN assay kit (Denka Seiken, Japan). Creatinine in rat blood was measured using a commercial kit (Denka Seiken, Japan). K^+/Na^+ assay kit (Toshiba, Japan) served to determine urine excretion of K^+ and Na^+.

5 METHODS

5.1 PREPARATION PROCESS OF FACTOR H

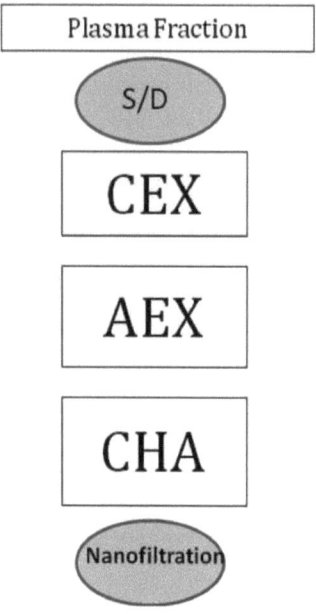

Figure 3 - CFH preparation process
The process comprises an initial solvent/detergent (S/D) treatment of the starting material (plasma fraction) followed by cation exchange chromatography (CEX), anion exchange chromatography (AEX) and ceramic hydroxyapatite chromatography (CHA) and a final nanofiltration step (described in Brandstaetter et al, 2012).

CFH was purified from pooled human plasma donations (human plasma fractions) according to the method by Brandstaetter et al. (2012). The individual steps of the purification process are briefly depicted in Fig. 3.

5.1.1 Biochemical and Functional Characterization of Intermediates and purified, final CFH

Biochemical and functional investigations of intermediates and purified, final CFH were described in detail in Brandstaetter et al. (2012).

In brief, functional characterization of purified, final CFH included testing for cofactor activity (CFH being essential cofactor for factor I-mediated C3b cleavage), decay acceleration activity (CFH competition with factor B fragment Bb for binding to C3b to promote the decay of C3 convertase complexes) and for protection of sheep erythrocytes from complement-mediated hemolysis.

5.2 RAT MODEL OF INTESTINAL IRI

5.2.1 Ethics Declaration

All animal procedures were conducted in the laboratories of the Austrian Institute of Technology (AIT), Health and Environment Department in 2444 Seibersdorf. Permission for performing the animal experiments was obtained from the provincial government of Lower Austria (LF1-TVG-41/006-2012).

5.2.2 Animals

Test species/ strain	Rat, Him: OFA (Sprague Dawley)
Breeder	Medical University of Vienna Division for Laboratory Animal Science and Genetics, A-2325 Himberg, Brauhausgasse 34
Number of groups	8 groups
Sex of animals	Female
Quantity, individual	Group 1: 4 rats (No. 1-4) Untreated

numbers and groups of animals	Group 2:	4 rats (No. 11-14)	Anesthesia
	Group 3:	4 rats (No. 21-24)	Sham
	Group 4:	4 rats (No. 31-34)	IRI
	Group 5:	3 rats (No.51-53)	Anesthesia + CFH (dose 1)
	Group 6:	6 rats (No.61-66, 41)	IRI + CFH (dose 1)
	Group 7:	3 rats (No.71-73)	IRI + CFH (dose 2)
	Group 8:	3 rats (No.81-83)	Sham + CFH (dose 1)
Body weight (at administration)	200 – 300 g		
Animal Identification	Felt tip pen on the root of the tail		
Hygiene	Optimal hygienic conditions		
Room temperature	About 22°C ± 2°C (continuous monitoring and recording)		
Relative humidity	About 40 % - 70 % (continuous monitoring and recording)		
Lighting	Only artificial light from 6.00 a.m. to 6.00 p.m.		
Cages	Group caging. Makrolon cages type III, (42 cm long, 26.5 cm wide and 15.5 cm high)		
Food	Ssniff-maintenance diet for rats and mice R/M-H (item V1534-300, Ssniff Spezialdiäten GmbH, 59494 Soest, Germany), autoclavable, *ad libitum*		
Water	Tap water from water bottles, *ad libitum*. Random samples of the water are analysed for contaminants and germ content by the Austrian Agency for Health and Food Safety (AGES), Spargelfeldstraße 191, A-1226 Vienna. The limits of tolerance are identical with that used for drinking water for humans in Austria (with the exception of the additional acidification).		
Bedding material	Aspen wood chips, ABEDD®, LAB & VET Service GmbH, Hasnerstraße 84/6, 1160 Vienna; germ reduction by autoclaving; changed 1/week. Random samples of the bedding material are analyzed for contaminants by the supplier.		
Environmental Enrichment	A "rat tube" (red polycarbonate shelter, Ø 8 cm, 15 cm length) is offered per cage. Germ reduction by autoclaving.		
Acclimatisation	7 days		

5.2.3 Surgical procedure

NO TREATMENT

Group 1: Untreated

Rats of group 1 were anesthetized by isofluran inhalation, additional analgesia was gained via carprofen intramusculary (4 mg/kg). After confirmation of death by a veterinarian, tissue dissection was performed as described under 5.2.5 "Tissue dissection".

ANESTHESIA AND NO SURGERY

Group 2: Anesthesia

Group 5: Anesthesia + CFH (dose 1)

Rats of both group 2 and 5 got anesthetized via isofluran inhalation for 90 min. Additional analgesia was done by carprofen intramusculary (4 mg/kg). During the anesthesia, electrocardiogram and body temperature were continuously monitored with a Small Animal Monitoring and Gating System (see 4.2 "Materials for the rat model"). In difference to group 2 (anesthesia only), rats of group 5 were injected with human CFH (see 5.2.6 "Administration of purified human CFH") immediately after onset of anesthesia. After 90 min anesthesia, rats were euthanized by decapitation, and tissue dissection was performed as described (see 5.2.5) after confirmation of death by a veterinarian.

SHAM SURGERY

Group 3: Sham

Group 8: Sham + CFH (dose 1)

Sham-operated rats of groups 3 and 8 were entirely treated like groups of IRI surgery and hence were undergoing a surgical procedure with the only difference that the arterial and collateral blood vessels of the terminal ileum were not ligated, thus creating sham surgery. Group 8, in addition, received human CFH at dose 1 prior to surgery (see 5.2.6 "Administration of purified human CFH").

IRI SURGERY

Group 4: IRI

Group 6: IRI + CFH (dose 1)

Group 7: IRI + CFH (dose 2)

Rats of group 4 (IRI), group 6 (IRI + CFH dose 1) and group 7 (IRI + CFH dose 2) underwent a surgical procedure. In group 6 and 7, animals were administered human CFH of dose 1 (group 6, 700 µg human CFH per ml plasma) or dose 2 (group 7, 350 µg human CFH per ml rat plasma) as described below (see 5.2.6 "Administration of purified human CFH"). Rats got anesthetized via isoflurane inhalation. Additional analgesia was obtained by carprofen intramusculary (4 mg/kg). During the anesthesia, electrocardiogram and body temperature were continuously monitored with a Small Animal Monitoring and Gating System (see 4.2 "Materials for the rat model"). Throughout the procedures, rats were placed on a heating table (+37°C). After laparatomy, exteriorization of a segment of the ileum was determined by visual inspection. This segment should be approximately 2 cm proximal to the caecum. Next, its vascular supply was identified. The collateral vessels were ligated by sutures and the supplying mesenterial vessel of the segment was ligated via a microvascular clamp. Fig. 4 shows the positions of the vascular clamp (clamp) and the sutures (ligatures).

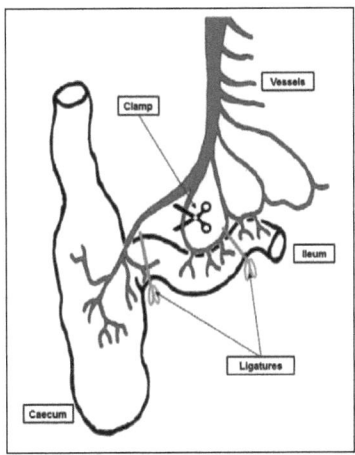

Figure 4 - Positioning of vascular clamp and sutures

Five mm distal and proximal from the applied ligatures (Fig. 4), two-colored filaments were attached at the anti-mesenteric side of the intestine to mark the entire length of ischemic intestine and especially know to the orientation of proximal/distal along the ischemic intestine when dissecting tissues for histology (see 5.2.7 "Histology and Light Microscopy"). This enabled first direct histological comparability of intestinal injury of ileum among all study groups and also allowed assessment whether the graded intestinal injury (see 5.2.8 "Pathology and Villi Length Analysis") might not be the same and hence vary along the entire length of the ischemic ileum. Following the ligation, the exposed intestine was then gently relocated into the abdominal cavity including the microvascular clamps. The abdominal wall was closed to create warm ischemia for a 30 min time period. After this the abdominal cavity was reopened, and the vascular clamp and the sutures were removed. The abdominal wall was closed for a second time tightly. Reperfusion was allowed for a 60 min period. Then rats were euthanized by decapitation. Subsequently, individual samples of the terminal ileum, the duodenum, the jejunum, the colon, the pancreas, the liver, kidneys, the lungs, and the brain were obtained in fixation medium as described in detail below (see 5.2.5 "Tissue dissection").

5.2.4 Blood sampling

- Group 1: Untreated

Rats were bled immediately after the onset of anesthesia (baseline) and subsequently sacrificed.

In groups not treated with human CFH, the following blood sampling was performed:

- Group 2: Anesthesia
- Group 3: Sham
- Group 4: IRI

The animals were bled retro-orbitally at beginning of anesthesia (baseline), after 30 min by puncture of the *vena saphena* and finally before euthanasia after 90 min anesthesia (terminal bleeding). At each timepoint 60-100 µL Lithium-heparin plasma was obtained and aliquots of 10 µL were immediately shock frozen in liquid nitrogen and stored at -70°C for subsequent analysis (see 5.2.11 *"Rat blood biochemistry"*).

In all groups treated with human CFH, the following additional blood sampling was done:

- Group 5: Anesthesia + CFH (dose 1)
- Group 6: IRI + CFH (dose 1)
- Group 7: IRI + CFH (dose 2)
- Group 8: Sham + CFH (dose 1)

200 μL of K_3-EDTA-anti-coagulated whole blood was additionally obtained at baseline and at terminal bleeding. These samples were immediately stored on ice and were handed over to an external laboratory (In Vitro-Labor GmbH, Rennweg 95, A-1030 Vienna) for determination of hematologic parameters (see 5.2.11.7 "Hematology").

5.2.5 Tissue dissection

Dissection support was provided by an experienced histopathology technician of the AIT. The order of organ retrieval was timed and maintained for each study animal and all dissected tissue pieces were of equal size comprising 5 x 5 mm. Individual samples of the duodenum, the jejunum, the terminal ileum, the colon, the pancreas, the liver, kidneys, the lungs, and the brain were obtained in this order from each study subject. Immediately after dissection, each tissue was fixed in cold phosphate-buffered formalin (+4 – 8 °C) for 8 hours at +4°C. First, the small intestine was exteriorized, caecum and stomach were removed and a duodenal and a jejunal segment were harvested. These served as internal controls regarding histo-pathologic scoring and villi length analysis (see 5.2.8). Then, the filament-marked ileal segment was harvested. Intestinal content of all segments harvested was flushed by

0.9 % saline from proximal to distal. Two pieces of each intestinal segment (ileum, jejunum and duodenum) were dissected in the same orientation. The larger and distal piece was fixed in 10 % formalin buffered saline (FBS), the second, smaller and proximal tissue piece was snap frozen in an upright position into optimal cutting temperature (OCT) compound. Then, a flushed segment of the colon and the pancreas were harvested and fixed in formalin and OCT. Next, the large liver lobe and the right entire kidney were harvested, sectioned into two equally sized sagittal tissue sections (5 x 5 mm) and fixed in cold formalin and in OCT. Then, the entire lung was retrieved and the right lung sectioned into two equally sized pieces that were fixed in cold formalin and in OCT. At last, the large brain was carefully obtained as a whole, partitioned into the two hemispheres. The right large brain hemisphere was fixed in cold formalin while the left large brain hemisphere was snap-frozen embedded in OCT.

5.2.6 Administration of purified human CFH

Purified human plasma CFH was warmed up at 37°C for 15 min in a waterbath prior to intravenous (*i.v.*) injection into rats of various groups. Two doses of human CFH were administered and were between 2-4 fold higher than the endogenous rat CFH plasma level. Administration of human CFH of dose 1 resulted in a four-fold higher level resulting in approximately 700 µg/ml human CFH in rat plasma (dose 1) or two-fold higher level of approximately 350 µg/ml human CFH in rat plasma (dose 2). The injected amounts of human CFH corresponded to functional (hemoprotective) amounts of CFH determined by hemolysis assay (see 1.1 Hemoprotection). The chosen dose was based on published studies, in which mice were treated with purified human CFH (Fakhouri et al, 2010b; Griffiths et al, 2009).

5.2.7 Histology and Light Microscopy

After 8 hours of fixation of dissected rat tissue in 10 % formalin-buffered saline (FBS), buffers were changed with cold PBS buffer at +4°C three times before dehydrated by a series of graded ethanol concentrations using an automated dehydration machine (Leica) and embedded into paraffin (39601006, Surgipath Parablast, Leica Biosystems) on an embedding machine (Leica EG1150-H, Leica Biosystems) by standard procedures. Importantly, as the orientation from proximal to distal was marked by colored filaments (see "Tissue dissection"), the same orientation for embedding in paraffin was used. Sectioning of formalin-fixed paraffin-embedded (FFPE) tissues was performed using a manual rotary microtome. Light microscopy was done on a Zeiss Axio Scope microscope (Carl Zeiss Microscopy GmbH, Jena, Germany) and microscopic photographs (TIFF images) taken by a connected camera using AxioVision software version 4.0 (Carl Zeiss) for subsequent pathologic and morphometric evaluations. Light microscopy was also conducted on an Olympus AX70-TRF microscope using the Metamorph software for digital imaging (as TIFF file format) of tissues.

5.2.8 Pathology and Villi Length Analysis

Sections of formalin-fixed paraffin-embedded (FFPE) intestinal tissues were prepared on a manual rotary microtome (4 µm), stained by hematoxylin and eosin (H+E) due to standard protocols and microscopic photographs of known pixel-to-µm ratio were taken by means of the AxioVision 4.0 software connected to the camera of the Zeiss Axio Scope microscope. Villi length in calibrated microscopic photographs was determined using the software ImageJ (NIH, Bethesda, USA) and villi length was expressed in µm. Mucosal injury on individual villi was graded on a six-tiered scale according to the method of Chiu (Chiu et al, 1970). Only well-oriented villi longitudinally sectioned and with a clear connection between crypt and villus were included for scoring. Fig. 5 delineates the five-point Chiu scoring on behalf of H+E-stained paraffin sections of rat intestines from various study groups of this work.

Scores include normal appearing villi (grade 0), the development of subtle apical subepithelial space at the villi tips (grade 1), the formation of Gruenhagen spaces as extended epithelial lifting (grade 2), massive epithelial lifting with partial destruction of epithelial layer (grade 3), progressing denudation with exposed lamina propria and progressive loss of epithelium (grade 4) and complete disintegration of lamina propria and loss of epithelium, ulceration reaching the crypt area and significantly shortened villi heights including complete loss of villi (grade 5).

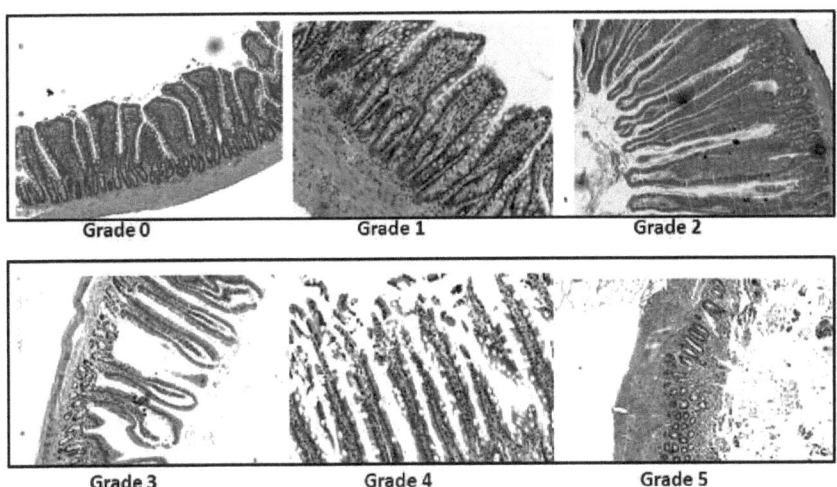

Figure 5 - Grading of Intestinal Injury on behalf of the mucosal

FFPE sections (3 µm) of kidney, liver, lung and brain tissues were prepared on a manual rotary microtome, stained by H+E and were evaluated by an external pathologist.

5.2.9 Immunohistochemistry

FFPE sections of rat intestines (4 µm) were de-waxed and re-hydrated by xylene followed by a series of graded ethanols. Heat-mediated antigen retrieval in 10 mM citrate buffer, pH 6 (target retrieval solution, Dako, S1699) was done during 20 min at +98°C followed by further 20 min of cool down of sections in citrate buffer (Dako, S1699). Sections were marked by a hydrophobic pen (Dako, S2002) and blocked for 30 min at RT in Sorensen phosphate buffer (PB), pH 7.2 containing 2 % rabbit serum (Dako, X090210), 1 % BSA (fraction V, Sigma), 0.1 % cold fish skin gelatin (Sigma, G7765), 0.05 % sodium azide (Sigma, S2002) and 0.05 % Triton X-100 (Sigma, T8294). Biotin-streptavidin blocking

(Dako, X0590) was performed and incubation with primary antibodies overnight at +4°C. Rat CFH was immunolocalized by polyclonal goat-anti-mouse/rat CFH IgG (sc-17951, Santa Cruz Biotechnology, California, USA).

Human CFH was immunolocalized by polyclonal goat-anti-human CFH IgG (AF4779, R&D Systems). Purified goat IgG from goat whole serum (> 80 %, I9140, Sigma) served as isotype control. All antibodies were diluted 1:200 in Sorensen PB buffer (pH 7.2) containing 1 % BSA, 0.1 % cold fish skin gelatin and 0.05 % sodium azide. Next, sections were incubated with a secondary biotinylated antibody rabbit-anti-goat IgG (affinity-isolated, Dako, E0466) diluted 1:1000 in wash buffer PB, pH 7.2 plus 0.5 % Tween-20 (Sigma, P1379) followed by incubation with streptavidin-peroxidase polymer (ultrasensitive, K3466, Sigma) diluted 1:1000 in washing buffer. Peroxidase-antibody complexes were detected with diaminobenzidine (DAB, Dako, K3466). Counterstaining of sections was done with Mayer´s Hematoxylin solution (Sigma, MHS16-500ml). Finally, slides were dehydrated through a graded ethanol series, cleared by xylene and mounted with Entellan mounting media (Merck, 1.07961.0100). Microscopy of IHC slides was performed on a Zeiss Axio Scope microscope (Carl Zeiss Microscopy GmbH, Jena, Germany) and microscopic photographs (TIFF images) taken by a connected camera using AxioVision software version 4.0 (Carl Zeiss). Additionally, an Olympus AX70-TRF microscope using the Metamorph software for digital imaging (as TIFF file format) of tissues was used.

5.2.10 Immunofluorescence Microscopy

Frozen sections from OCT-embedded snap-frozen rat intestinal tissue were prepared on a cryostat (8 µm sections, Microm HM550), dried at 37°C for one hour and stored at -80°C until analysis. Sections were brought to room temperature for 30 min, fixed in ice cold acetone during 5 min and were then air dried during another 30 min. Sections were washed 1x in PBS, 1x in PBS + 0.05 % Tween-20 (Washing buffer) and blocked during 30 min with PBS, pH 7.4 + 0.05 % Tween-20 + 0.1 % Triton X-100 + 5 % BSA. Slides were incubated overnight at +4°C in a moist and dark atmosphere of a slide staining tray with rabbit anti-human C3c antibody conjugated to fluoresceinisothiocyanate (FITC) (Abcam, ab4212) or rabbit anti rat IgG antibody – FITC conjugated (Sigma, F4512) as isotype control or only using antibody dilution buffer (PBS, pH 7.4 + 0.05 % Tween-20 + 1 % BSA). Antibodies were diluted 1:200 in the antibody dilution buffer and 100 µl were applied to the tissues. Nuclei of intestinal tissue on sections were then counterstained with the fluorescent dye Hoechst 33258 (Sigma, 861405), mounted by Fluoroprep mounting media (75521, bioMérieux) and permanently sealed by nail polish. Sealed slides were kept at 4°C chilled and dark. Fluorescence microscopy was performed on an Olympus Provius AX70TRF microscope connected to a xenon arc lamp and the software Metamorph for digital imaging. Software settings for photographing, fluorescence image contrast thresholds and exposure times were strictly maintained for all sections analyzed.

5.2.11 Rat blood biochemistry

5.2.11.1 Factor H activity – Hemolysis Assay

This assay was adapted from the version described above ("Hemoprotection"). Sheep erythrocytes were prepared from stabilized sheep blood (ORAW31, Siemens Healthcare, Marburg, Germany), washed 4 times in 10 ml cold Alsever´s solution (A3551, Sigma), gently mixed and precipitated by centrifugation at 1000-xg for 5 min at +4°C. A 33 % sheep red blood cell (SRBC) suspension in cold Alsever´s solution was prepared, of which $2*10^9$ SRBC per ml were used per assay. A stock solution of 700 µg purified human CFH per ml of rat plasma was prepared using the same purified human CFH *in vitro* as administered to the animals *in vivo*. Therefore, the stock solution comprised the same concentration of purified human CFH as administered to IRI rats. To generate a standard curve, the stock solution was diluted with pre-immune rat plasma in the same way as human CFH-containing rat plasma samples were diluted. The regression formula from the standard curve was used to recalculate the concentration of human CFH activity of each rat plasma dilution tested. 10 µl of human CFH-rat plasma dilutions were mixed with 10 µl of SRBC suspension in an equal reaction volume filled with PBS, pH 7.4, and pre-incubated for 30 min at +37°C under mild agitation (Heidolph Titramax 1000). Then, PBS, pH 7.4 containing 10 mM EGTA and 7 mM $MgCl_2$ were administered to start alternative complement-mediated hemolysis. Samples were incubated for another 30 min at +37°C and the reaction was stopped by immediate addition of ice cold PBS, pH 7.4 containing 20 mM EDTA. Lysed cells were pelleted by centrifugation (20.000-xg, 5 min, RT) and supernatants tested immediately for released haemoglobin by absorbance measurement at

414 nm (Molecular Devices photometer, Versamax). The assay enabled the determination of hemo-protective CFH in a linear concentration range between 0 and 70 µg/ml. Dilution of human CFH-containing rat plasma samples in order to meet this standard range was done using pre-immune rat plasma.

5.2.11.2 Factor H recovery in rat plasma – ELISA

Human CFH antigen concentration of plasma of all rats injected with human CFH in this study was determined by a commercially available ELISA (HK342, Hycult Biotech, Uden, The Netherlands). The ELISA did not cross-react to neither mouse nor rat CFH and indicated complete recovery of spiked human CFH into mouse or rat plasma (data not shown). Testing the administered human CFH preparation in four independent assays showed a mean inter-assay variance of 10.8 %. The ELISA was thus suited to specifically assess the recovery of human CFH in rat plasma over time. A pooled mean inter-assay variance of 18.4 % of human CFH antigen in plasma samples of rats of all study groups was found in three independent determinations. Rat plasma samples containing human CFH were diluted 1:5.000 and 1:10.000 in order to meet the linear range of the standard curve. The entire ELISA procedure was done as described by the manufacturer´s protocol.

5.2.11.3 Complement C3 activation status – Western Blot

Rat plasma samples were thawed and incubated in reducing and non-reducing sodium dodecylsulfate (SDS) sample buffer. In order to minimize post-thawing C3 activation, the time of processing of rat plasma until gel loading was kept at a minimum. Immediately after thawing (2 min), rat plasma samples were brought to 10 mM EDTA. SDS-PAGE was performed on pre-cast 4-20% TRIS-glycine gradient gels (EC6025BOX, Invitrogen) followed by transfer to nitrocellulose (Hybond C extra, GE Healthcare). Membranes were blocked (Superblock, Pierce) and incubated with a goat polyclonal antiserum to rat C3 (55713, Cappel) diluted 1:5.000 in washing buffer TRIS-buffered saline, pH 7.2 plus 0.5 % Tween-20 (Sigma) for 60 min followed by 30 min incubation with an anti-goat IgG-HRP Fc-fragment specific antibody (Calbiochem) diluted 1:10000 in washing buffer. Positive bands were detected by chemiluminescence with a luminol-based substrate (Super Signal, Pierce) on the LAS-3000 imaging station (LAS-3000, Fujifilm).

5.2.11.4 Complement Hemolytic Activity in plasma – CH50 Assay

Sheep erythrocytes were prepared from stabilized sheep blood (ORAW315, Siemens Healthcare), washed 4 times in cold Alsever's solution (A3551, Sigma), gently mixed and precipitated by centrifugation at 1000 xg for 5 min at 4°C. A 10 % (w/v) sheep red blood cell (SRBC) suspension in cold Alsever's solution was prepared. Upon drop-wise addition of 2 % (v/v) rabbit anti SRBC antibody (Amboceptor, ORLC25, Siemens Healthcare Diagnostics, 35041 Marburg, Germany) SRBC were sensitized during 30 min at 37°C in a waterbath. The optimal sensitization of SRBC was performed according to a standardized operation procedure (SOP) of Octapharma PPGmbH Vienna (000SOP026/11 – Determination of the anti-complementary activity of immunoglobulins due to European Pharmacopeia). On each experimental day, all rat plasma samples of this study were assessed for CH50 with a freshly prepared SRBC solution. All rat plasma aliquots were tested individually three times. Rat plasma was diluted 1:8, 1:16, 1:32, 1:64 and 1:128 in starting buffer (PBS, pH 7.4 containing 1 mM magnesium chloride and 0.5 mM calcium chloride) (Costabile, 2010). 40 µl of each plasma dilution was mixed with 160 µl sensitized SRBC solution and samples were incubated for 30 min at 37°C under gentle shaking (Heidolph Titramax 1000). The reaction was stopped by immediate addition of 100 µl ice cold PBS, pH 7.4 containing 10 mM EDTA. Lysed cells were pelleted by centrifugation (1.500-xg, 5 min, RT) and supernatants were tested immediately for released haemoglobin by absorbance measurement at 540 nm using a spectrophotometer (Molecular Devices, Versamax).

5.2.11.5 Complement C3 concentration of rat plasma – ELISA

The concentration of native C3 in rat plasma was determined by a commercial ELISA (IRTC3KT, Innovative Research). The inter-assay variance found after three independent experiments was 7.7 % (n = 3 assays).

5.2.11.6 Total protein of rat plasma – Bradford Assay

Protein concentration of rat plasma samples was determined by the Bradford Assay (Bradford, 1976) using a Coomassie reagent (23238, Coomassie Plus). Bovine serum albumin was used as protein standard (23209, Pierce). The inter-assay variance of samples was determined to be 3.5 % (n = 3 assays), however, the intra-group variance of protein values was higher despite manual constancy in retro-orbital bleeding during the animal studies and was up to 5.4 %.

5.2.11.7 Hematology

After onset as well as after 90 min of anesthesia, 200 µl of whole blood anti-coagulated in K-EDTA tubes was obtained, immediately stored and transferred to an external laboratory (INVITRO-Labor für veterinärmedizinische Diagnostik und Hygiene GmbH, Rennweg 95, A-1030 Vienna, www.invitro.at) for differential blood cell count analyses.

5.2.12 CFH expression in Caco-2 cell lysates – Western Blot

Caco-2 whole cell lysates (Abcam, ab-3950) were separated by SDS-PAGE and CFH expression determined by Western blotting using a polyclonal antiserum to human CFH (Comp Tech, A337). 20 µg protein of Caco-2 whole cell lysates per lane was separated by electrophoresis using 4-20 % TRIS-glycine pre-cast gels (Invitrogen). Transfer on nitrocellulose (Hybond C extra, GE Healthcare) following SDS-PAGE was done by wet tank system (XCell-II Blot Module, Invitrogen). Membrane was blocked (Superblock, Pierce) and incubated with a goat polyclonal antiserum to human CFH (Comp Tech, A337) for 60 min followed by 30 min incubation with an anti-goat IgG-HRP Fc-fragment specific antibody (401504, Calbiochem). Chemiluminescence was detected after incubation with luminol-based substrate (34075, Super

Signal West Dura, Pierce) using the LAS-3000 imaging station (LAS-3000, Fujifilm, Tokyo, Japan).

5.2.13 Statistical Evaluation

Comparison of mean Chiu scores and villi lengths between study groups was performed by analysis of variance (ANOVA) using the Tukey-Kramer method. A p-value < 0.05 was considered statistically significant. Comparison of differences of mean complement activity (CH50) and C3 concentrations in rat plasma obtained at baseline and after 90 min was determined by ANOVA using the Tukey-Kramer method on the alpha 0.05 significance level. All data were analyzed using SAS software, version 9.3 (SAS Institute Inc., Cary, USA).

5.3 Rat model of renal IRI

Administration of purified human CFH in an established rat model of renal IRI was conducted in the laboratories of a clinical research organization in Taiwan (Eurofins PanLabs formerly Ricerca Biosciences LLC, Bothell, WA 98021, USA, Pharmatoxicology division). The rat model of renal IRI was conducted as described previously (Chintala et al, 1994; Nakamoto et al, 1987; Tripatara et al, 2007). Briefly, male rats of the Sprague Dawley strain were anesthetized by pentobarbital (50 mg/kg, intraperitoneal) after receiving hirudin (Refludan®, 0.2 mg/kg, *i.v.*). Five minutes before laparatomy, rats received an *i.v.* injection of dose 1 (700 µg/ml) human CFH or vehicle (PBS). Bilateral renal artery occlusion was performed for 45 min using microvascular clamps followed by a reperfusion phase of 2 days before euthanasia and tissue retrieval in the IRI groups. A third group of sham-operated rats was included in this study that underwent the same surgical procedure except renal artery occlusion. Tissue dissection was conducted under the same strict modalities as in the intestinal model. The following outcome measures were evaluated: mortality, body weight, kidney weight, urine volumes, urine chemistry (creatinine, Na^+, K^+), blood chemistry (creatinine, blood urea nitrogen (BUN), Na^+, K^+), fractional excretion of Na^+ (FE_{Na}), endogenous creatinine clearance, content of Kidney Injury Molecule-1 (KIM-1) and N-acetyl-β-D-glucosaminidase (NAG) in urine, and histopathology. Histo-pathological analysis of left and right kidney were erformed by grading of lesion severity according to a previous report (Shackelford et al, 2002).

6 RESULTS

6.1 Native, functional CFH was purified from human plasma fractions

A fully functional concentrate of human CFH was purified from CFH-enriched human plasma fractions as described by Brandstaetter et al, 2012. The new process gave rise to a 94% pure CFH preparation and produced highly purified CFH at laboratory scale at amounts that sufficed pre-clinical testing in small animal models like rodents.

Due to Brandstaetter et al, 2012, the purified CFH preparation demonstrated

- efficient and dose-dependent cofactor activity for complement factor I-mediated cleavage of C3b in a fluid phase assay
- dose-dependent decay of nickel-chloride - stabilized C3 convertase complexes in an ELISA-based assay
- protection of sheep erythrocytes from human serum complement-mediated hemolysis in a dose-dependent manner

In summary, a functional CFH concentrate could be gained in a novel purification process from human plasma fractions of industrial plasma fractionation by three conventional chromatographic steps.

As a novelty to previously published CFH isolation processes, truncated and thus dysfunctional CFH species could specifically be removed by this purification process. This led to a fully functional and native preparation of purified human CFH (Brandstaetter et al, 2012).

6.2 A rat model of intestinal IRI induces local mucosal tissue injury and complement deposition

In order to study the value of CFH therapy in a preclinical model of complement activation, a rat model of intestinal IRI was established in the laboratories of the Austrian Institute of Technology (Health and Environment Department, A-2444 Seibersdorf). The influence of complement to pathology of IRI has been extensively demonstrated in the last decade in mice (Austen et al, 2003; Lee et al, 2010; Lu et al, 2008), rats (Eror et al, 1999; Hill et al, 1992; Padilla et al, 2007; Souza et al, 2005; Wada et al, 2001) and humans (Grootjans et al, 2010; Matthijsen et al, 2009). A CFH-based complement inhibitor protected mice from intestinal IRI without clarifying the underlying mechanism (Huang et al, 2008a). For these reasons a rat model of intestinal IRI was established.

An illustration on a time scale of individual steps performed is depicted in Fig. 6. Female Sprague Dawley rats were subjected to 30 min warm and complete ischemia induced in the terminal ileum followed by 60 min of reperfusion.

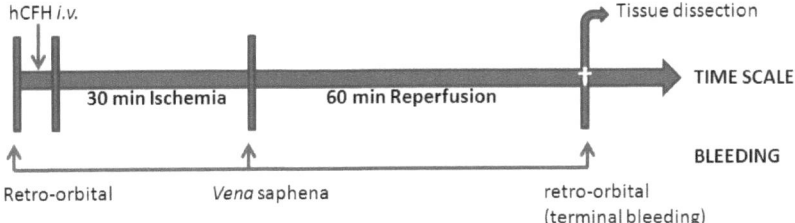

Figure 6 - Intestinal IRI in rats on a time scale.
Complete and warm Ischemia was allowed in terminal ileum of rats during 30 min that was followed by a 60 min reperfusion phase. Five minutes prior to laparatomy, human CFH (hCFH) was injected intravenously (*i.v.*) into a subset of animals. Rats were sacrificed (†) by decapitation and tissues were dissected as described in the Method's section. Rats were bled at time points indicated by bleeding of the retro-orbital sinus or vena saphena to obtain 100 µl of plasma. After euthanasia, animals were terminally bled from the retro-orbital sinus and the maximal amount of whole blood and plasma harvested.

Complete ischemia was assured by ligation of both the arterial blood supply with a microvascular clamp and simultaneously of the collateral blood supply by suture filaments applied on both ends of the targeted branch of the terminal ileum (illustrated in Fig. 4).

Rats were bled retro-orbitally at baseline, after 30 min ischemia via *vena saphena* and after 90 min (post 60 min reperfusion) at the retro-orbital sinus. Blood was drawn from *vena* saphena after 30 min ischemia as animals could not be separated from the isoflurane inhalation mask to access the retro-orbital sinus. In each case, lithium-heparin plasma was immediately gained and stored at -70°C until analysis. In Fig. 7 visualizations of the surgery of the experimental model are provided.

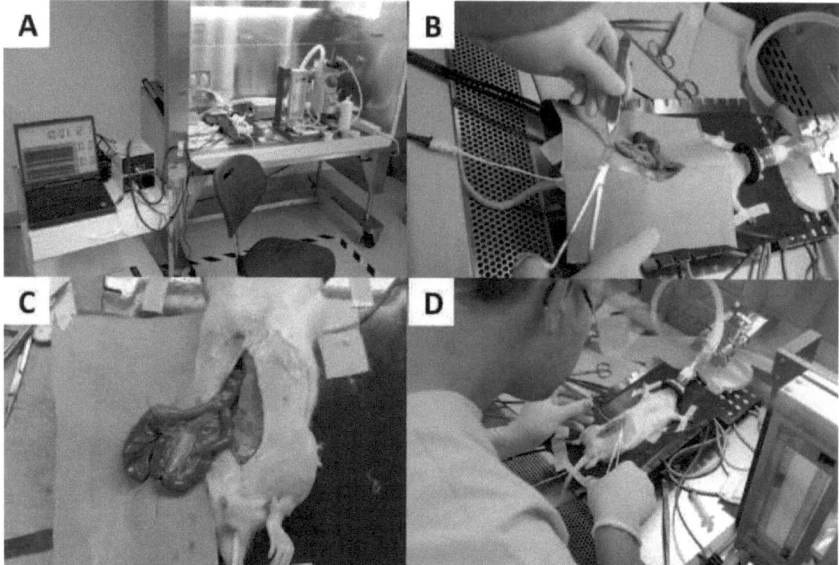

Figure 7 - Surgery to create intestinal IRI in rats.
(A) Vital monitoring of heart beat, respiratory rate and body temperature of rats anesthetized by isoflurane inhalation was performed. Rats were placed on a heating table under laminar flow.
(B) Exposure of the terminal part of small intestine to identify terminal ileal segment and its vascular supply. (C) Terminal ileum with applied filament ligations to suppress collateral blood supply.
(D) Veterinary surgeon tightly reclosing fascia and skin.

Ischemia was induced at the same segment of the terminal ileum. In addition, a topographically identical segment of the terminal ileum was removed from animals of all other study groups for comparability of mucosal injury grades among groups. Warm ischemia was assured as the entire small intestine including the microvascular clamps was relocated into the peritoneum after exterioration of the small intestine.

Fascia and skin were tightly closed in a two layered fashion before ischemia and before reperfusion. Every single rat - out of a total of

32 animals in this study –survived. In order to allow comparison of tissues of all study animals on the histological and histo-chemical level, animal and tissue handling were done in a reproducible manner, including reproducible surgical procedures and tissue dissections as well as the same modalities of tissue fixations.

GRADING OF INTESTINAL INJURY

Injury of the gut mucosa was evaluated on basis of a largely used scoring system according to Chiu (illustrated in Fig. 5). In addition, villi lengths were determined, as villous contraction is considered an early consequence of intestinal ischemia (Blikslager et al, 2007). The suitability of the rat model for therapeutic intervention required the ability to sensitively and significantly discriminate clinical readouts (mucosal tissue injury) between treated (IRI) and untreated (sham) rats. A clinically relevant improvement by any mono-therapeutic means in gut IRI, like CFH treatment solely, had to achieve an improvement of at least one Chiu score unit.

Provided the feasibility of surgical reproducibility in this model, normal distribution of the average biological injury score in the rat intestine was assumed. In order to establish the rat model, histo-pathologic grading of mucosa injury was performed in groups comprising anesthetized, sham-operated and IRI-operated rats. Histo-pathologic grading of mucosa injury was conducted along the entire gut. In addition, completely untreated rats (immediately sacrificed upon anesthesia) were included to investigate intestinal injury under physiological situation in comparison to that of manipulated animals.

Fig. 8 shows mean and SEM of Chiu scores and villi lengths for each group and intestinal segment. A comparison of mean Chiu scores and villi lengths of each intestinal segment between these study groups was conducted by one way ANOVA. Importantly, a significant difference of the mean Chiu score by more than one score unit was found when comparing IRI rats with sham-operated, anesthetized and untreated rats and was found statistically significant only when comparing the ileum. No significant difference in mean score was found in any other segment of small intestine between these study groups.

In addition, ileal villi of IRI rats were significantly smaller than those of sham-operated, anesthetized and untreated animals (Fig. 8B). Villi lengths of the Sprague Dawley strain used were in accordance with published data (Gardner & Steele, 1989).

The representative histological appearance of intestinal segments of sham and IRI rats was illustrated in Fig. 9. In comparison to sham controls, intestinal villi of IRI rats demonstrated severe mucosal injury that ranged from disintegrated and edematous villous stroma down to completely denuded villi (Fig. 9).

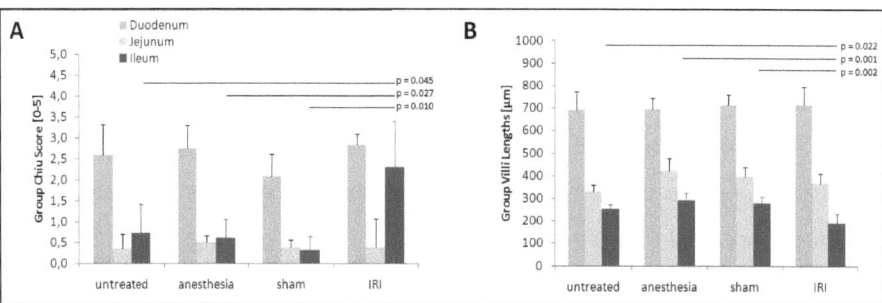

Figure 8 – Intestinal IRI but not sham surgery induces increased mucosal pathology and decreased villi heights in ileum Bars represent mean Chiu scores (A) and mean villi lengths (B). 30 to 40 villi per intestinal section per rat were scored according to Chiu and a mean score of 4 rats (a total of n=150 villi per group +/- SEM). Villi lengths were determined on microscopic photographs of H+E stained intestinal sections via ImageJ. Significant differences in mean score (A) and mean villi heights (B) of all intestinal segments between groups were analyzed by one way ANOVA (α=0.05, Tukey-Kramer method).

Villi of rats subjected to intestinal IRI demonstrate mucosal injury ranging from significantly smaller villi with partly deteriorated muscularis but rather intact epithelium with oedema and partly disintegrated villous stroma down to completely denuded villi with dilated vessels, disintegration of villous stroma, partly deteriorated crypts and outer muscularis and lastly with great loss of epithelium (Fig. 9).

Note that mucosal injury was induced solely in the ligated terminal ileum as scores of duodenal and jejunal villi of both sham and IRI rats were comparable to anesthetized and healthy rats (Fig. 8). Duodenum and jejunum were harvested from each animal and served as internal controls regarding pathology score and villi lengths. The decreasing villi lengths (and correspondingly decreasing scores) from proximal (duodenum) to distal (ileum) are according to gut physiology (Gardner & Steele, 1989) and reflect the known susceptibility of larger villi lengths to hypoxic stress (Blikslager et al, 2007; Vollmar & Menger, 2011).

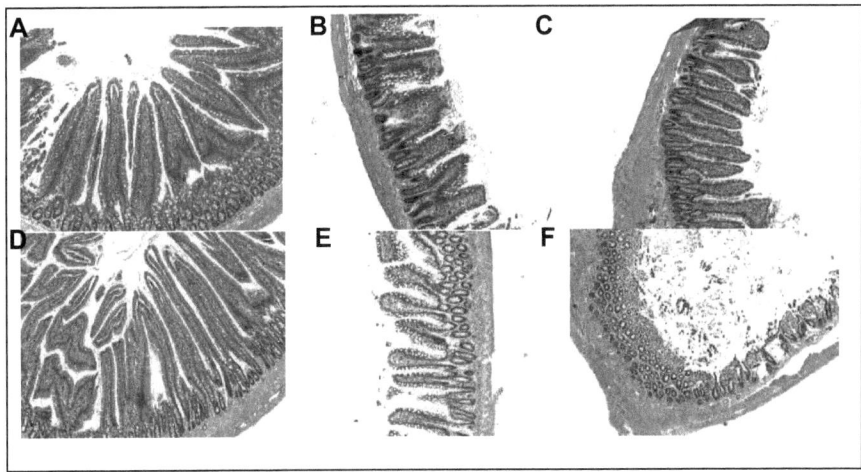

Figure 9 - Morphological appearance of rat intestines after sham surgery and intestinal IRI Representative 4 µm transverse sections of paraffin embedded (A, D) duodenum, (B, E) jejunum and (C, F) ileum of one representative rat subjected to sham surgery (A-C) or intestinal IRI (D-F) were stained by H+E and form the basis for subsequent Chiu scoring and determination of villi lengths.

6.2.1 No systemic complement activation induced in the rat model

Next, it was addressed whether complement activation took place in this rat model. Because no test systems are available to measure rat C3a-desArg, potential systemic complement activation was determined by Western blotting of rat plasma to detect any C3 activation fragments. The feasibility to investigate the C3 activation state in plasma of CFH-deficient mice by Western Blot has been recently published (Paixao-Cavalcante et al, 2009). The Western approach revealed the presence of a 43 kDa-C3-specific band, corresponding to iC3b, a result of complement activation (Fig. 10, A). This band showed mean increases of intensity between baseline and 90 min of 1.3 % (+/- 0.8 % SEM). However, there was no difference between any of the study groups.

6.2.2 Detection of C3 levels of rat plasma

As systemic complement activation would lead to significant changes of plasma C3 levels, concentrations of native C3 in rat plasma of all study groups were measured by rat C3-specific commercial ELISA (Innovative Research).

Detected baseline levels of plasma C3 were in line with the literature (Daha et al, 1979). Reductions of plasma C3 concentration between baseline and after 90 min were detected in all groups irrespective of human CFH administration (Fig. 10, C). No statistical significance was found regarding C3 concentrations of rat plasma obtained at baseline and at the 90 min time point when comparing all groups with each other.

6.2.3 Detection of Complement Hemolytic Activity (CH50) of rat plasma

To understand these data on a functional level, complement activity of rat plasma samples was determined using the classical complement hemolytic (CH50) assay. The readout of the CH50 assay is the serum or plasma dilution which leads to complement-mediated lysis of 50 % of antibody-sensitized sheep erythrocytes (Mayer, 1961). CH50 is largely but not exclusively affected by C3 as the most abundant complement component. CH50 at baseline was not statistically different in any group (Fig. 10, B). In contrast, IRI rats and all other study groups receiving human CFH demonstrated a reduction in CH50 after 90 min that was significant when compared to CH50 after 90 min of anesthetized rats and sham controls, but was not significant when compared to each other (Fig. 10, B).

6.2.4 Determination of total protein content of rat plasma samples

As a third measure to better understand these reductions of plasma C3 and CH50 levels, total protein levels of rat plasma drawn at baseline and after 90 min were determined by the Bradford assay (Bradford, 1976) and data are presented as percentual plasma protein difference between baseline and the 90 min time point (Fig. 10, D). Average baseline levels of total protein in rat plasma were consistently at around 50 mg/ml and were not statistically different between any study group. A correction of the CH50 values with the respective total plasma protein levels for each group was performed, but did not change the significant differences of CH50 between baseline and 90 min found in every group.

Administration of human CFH at dose 1 (700 µg human CFH per ml rat plasma) led to plasma volume dilution factor of about 7 %, while administration of human CFH at dose 2 (350 μg human CFH per ml rat plasma) led to a plasma volume dilution factor of about 3.5 %. The pattern of percentual plasma protein differences between baseline and 90 min reflected the plasma dilution factors between the study groups (Fig. 10, D). Thus, subtraction of the plasma expansion factor by human CFH injection from levels of percentual plasma protein differences between baseline and 90 min of CFH-treated groups resulted in levels found in the respective non-CFH-treated group. This finding explained that reductions of C3 concentrations and CH50 in plasma of CFH-treated animals were due to plasma expansion by human CFH administration. In addition, the percentual plasma protein difference between baseline and 90 min in the IRI group was similar to sham-treated animals (Fig. 10, D).

In summary, the reduced CH50, C3 and protein levels in plasma after 90 min of all CFH-treated groups were similar to the untreated IRI group and were a result of plasma expansion. These findings therefore indicate that IRI surgery did not lead to systemic complement activation in this rat model.

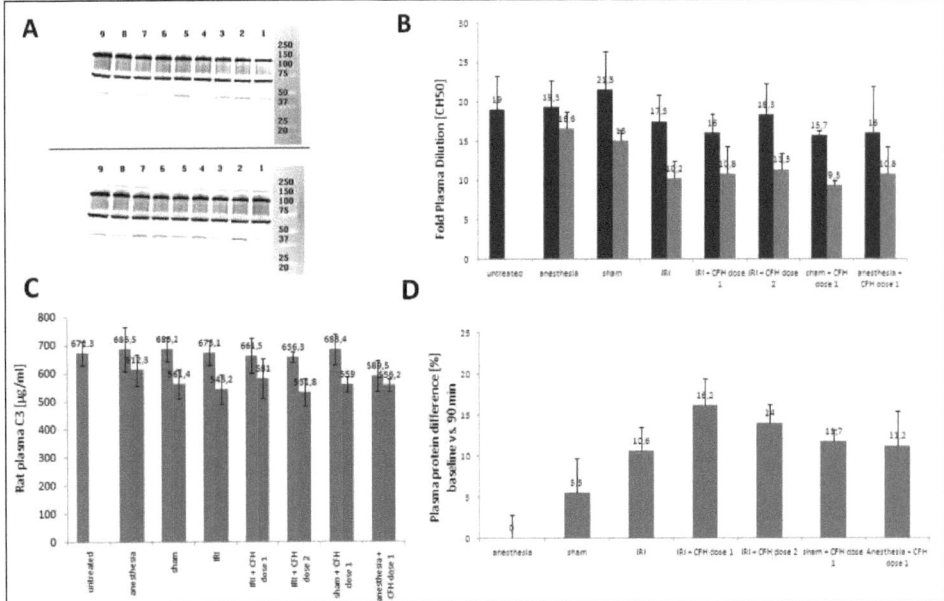

Figure 10 – Reduced plasma protein, C3 and CH50 in rats

(A) C3 activation state was tested by Western Blotting of plasma of sham operated rats (lanes 1-4), IRI rats (lanes 5-8) and pre-immune rat plasma (lane 9) under reducing conditions. Upper blot shows rat plasma drawn at baseline, lower blot shows plasma of the same rats drawn after 90 min. Membranes were probed for 60 min with goat anti rat C3 antibody (Cappel)

(B) Hemolytic Activity (CH50) of rat plasma. The average change in plasma dilution factor between baseline (dark blue) and 90 min (light blue) is shown (mean values per group +/- SEM of 3 experiments). Differences between groups regarding CH50 at baseline and at 90 min were analyzed by one way ANOVA (α=0.05, Tukey-Kramer)

(C) Rat C3 concentrations in plasma between baseline (blue) and 90 min (red) of all groups were determined by ELISA (Innovative Research, IRTC3KT) as mean concentrations in µg/ml per group +/- SEM of 2 experiments are given. Differences between groups regarding plasma C3 at baseline and at 90 min were analyzed by one way ANOVA (α=0.05, Tukey-Kramer)

(D) Percentual loss of total protein of rat plasma is presented as average of plasma protein loss in % per group (+/- SEM, n=3 experiments). Total plasma protein content was determined by Coomassie reagent and calibrated by bovine serum albumin (both Pierce).

6.3 Intestinal IRI induced local complement deposition in the ischemic gut

Previous reports showed that intestinal IRI elicited local inflammatory responses and local deposition of complement along the gut (Atkinson et al, 2005a; Huang et al, 2008a; Souza et al, 2005). Therefore, local complement deposition at rat ileum was analyzed by immune-fluorescence microscopy. Complement protein C3 was detected on frozen sections of the ileum of different groups using an antibody against human C3c conjugated to FITC (Abcam), which is cross-reactive to rat C3. This antibody was described to react with the C3c part of native C3 and C3b by the manufacturer.

Slight C3 staining was observed within vessels of mucosal stroma of anesthetized (Fig. 11, A) as well as sham-operated rats (Fig. 11, B). Occasionally, faint C3 staining could be seen at villi tips in ileum of these animals. Omission of primary antibody and application of an isotype control antibody (both directly conjugated to FITC) did not yield a positive signal (not shown).

In contrast, strong complement deposition on villi tips and of shedded cells in the gut lumen could be detected in ileum of rats subjected to intestinal IRI (Fig. 11, C&D, 2 different IRI rats). More specifically, C3 staining in IRI rats corresponded to detached and/or shedded epithelial cells and sub-epithelial capillaries with a strong staining in villi tips in agreement with previous reports (Fleming et al, 2010; Pope et al, 2012).

These results demonstrated the specific induction of complement activation locally in the intestine of rats undergoing intestinal IRI and are in agreement with an array of studies showing local complement deposition of the intestine in rodent IRI models (Atkinson et al, 2005a; Huang et al, 2008a; Souza et al, 2005).

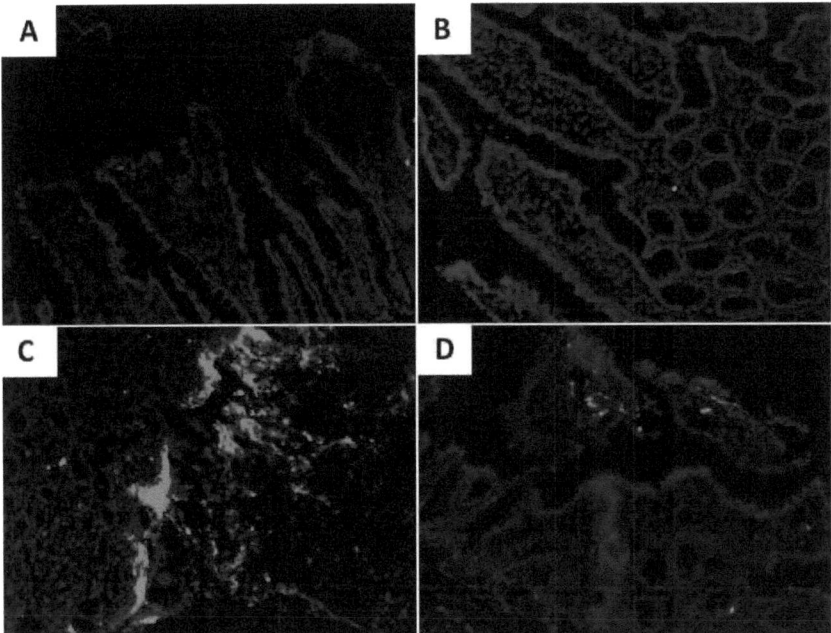

Figure 11 – Local C3 deposition at rat ileum subjected to IRI.
Representative microscopic photographs of immunefluorescence analysis of C3 on frozen sections of (A) anesthetized, (B) sham-operated and of (C, D) two different rats subjected to intestinal IRI are shown. Local C3 deposition was detected using a polyclonal rabbit anti human C3c antibody conjugated to FITC (Abcam, ab4212). Nuclei were counterstained by Hoechst (Sigma). No background fluorescence was detected neither by omission of C3c antibody nor by using polyclonal rabbit anti human IgG conjugated to FITC (Sigma, F4512, not shown). Equal microscopic magnifications (20x) and software settings were used.

In summary, a rat model of intestinal IRI was established with significant induction of mucosal injury and local complement deposition in ischemic ileum. The intestinal injury was induced locally in the terminal ileum by complete and warm ischemia and reperfusion. No remote organ injury in any peripheral tissue or other segment of small intestine was created. Furthermore, no indication for systemic complement activation was found. In contrast, a marked and specific induction of local complement deposition resulted from intestinal IRI in rats in agreement to previous publications.

6.4 Administration of human CFH does not significantly protect rats from intestinal IRI

Having established the rat model of intestinal IRI with local complement activation and injury of the gut mucosa, the effect of human CFH treatment on IRI was studied. Rats undergoing gut IRI were intravenously injected with two dosages of purified human CFH in order to identify a dose-dependent effect on mucosal injury.

The dosages were extrapolated from mouse models of human disease that were injected 0.5 mg of purified human CFH (Fakhouri et al, 2010b; Griffiths et al, 2009). Administration of 500 µg of human CFH to mice with CFH levels of ~ 0.5 mg/ml (Alexander & Quigg, 2007) results in a doubling of endogenous plasma levels. In light of this, it was chosen to double the endogenous plasma CFH level of rats (dose 2, 350 µg human CFH per ml rat plasma) and to include another treatment group increasing endogenous rat levels four-fold by human CFH injection (dose 1, 700 µg human CFH per ml rat plasma). In addition, the lower dose of human CFH was also administered to sham-operated and anesthetized rats to allow for comparison of all experimental groups treated with human CFH.

Administration of human CFH was performed five minutes prior to start of laparatomy. In total, four rats were subjected to intestinal IRI without treatment in this study. In two IRI rats no significantly elevated pathology was induced (Fig. 12, A, group 3, lower red dots) while in other two IRI rats a clinically relevant mucosal injury was found (Fig. 12, A, group 3, upper red dots).

The reason for the low injury in these two IRI rats was the manually applied attachment of filaments to restrict the collateral blood supply of the ileum, a major source of deviation in this rat model. However, the narrow distribution of injury scores of all IRI rats treated or not with human CFH combined (Fig. 12, A, groups 3-5) sustained the assumption of normally distributed injury among IRI rats. On average, the four IRI rats displayed significantly reduced villi lengths as result of IRI with an average of 187 µm compared to 277 µm in sham operated animals (Fig. 12, B, group 3).

Strikingly, the injection of purified human CFH did not significantly protect rats from mucosal injury after intestinal IRI (Fig. 12, A), nor did it have an effect on mean villi lengths (Fig. 12, B).

Neither of two administered doses of human CFH resulted in significant protection from gut IRI in rats and therefore did not prove the major hypothesis of this work.

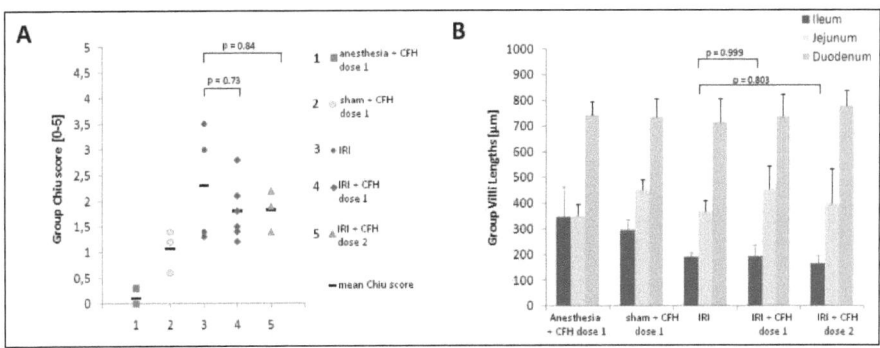

Figure 12 – Human CFH injection does not rescue rats from intestinal IRI.
(A) Averaged Chiu scores per experimental group and individual scores of each animal per group are presented in scattered blot.
(B) Averaged villi lengths per group comprising 3 to 6 rats per group are illustrated as bars (+/- SEM of an equal total number of villi per group). Differences of mean scores and of mean villi lengths were compared between groups by one way ANOVA (α=0.05, Tukey-Kramer).

6.4.1 Inflammatory cells did not significantly contribute to intestinal IRI

A growing body of evidence incriminates inflammatory blood cells and their activation products in the pathophysiology of intestinal IRI (Rodrigues & Granger, 2010). Depletion of resident macrophages or neutrophil priming by hypoxic preconditioning has been shown to protect rats from intestinal IRI (Chen et al, 2004; Lu et al, 2012).

Differential blood counting of leukocyte populations of whole blood drawn at baseline and after 90 min in groups treated with human CFH demonstrated no significant changes in circulating leukocytes between these time points (Table 3). Interestingly, IRI groups treated with human CFH at dose 1 and dose 2 showed a significant increase of neutrophils in peripheral blood after 90 min (Table 3).

However, careful histological examination revealed no significant infiltration of inflammatory cells in ileum or peripheral tissues like lungs, liver and kidneys in any of the study groups. Thus, it is unlikely that inflammatory cells contributed - at least to a large part - to the mucosal injury induced in this rat model.

	baseline Anesthesia + CFH dose 1			90 min Anesthesia + CFH dose 1			p =	baseline Sham + CFH dose 1			90 min Sham + CFH dose 1			p =
Rat #	51	52	53	51	52	53		81	82	83	81	82	83	
neutrophil leukocyte (%)	7,6	14,5	14,6	12,5	15,1	17,5	0,363	7	11,9	13,3	14,7	18,7	30	0,104
lymphocyte (%)	85,6	80,6	81,4	81,3	79,4	76,6	0,172	89,5	84,4	81,5	81	77,2	65,5	0,113
monocyte (%)	3,3	2,1	2	2,6	2,5	2,1	0,888	1,6	1,5	1,9	1,3	1,7	1,4	0,305
eosinophil leukocyte (%)	2,5	1,8	1,2	1,9	1,9	2,7	0,509	0,9	1,1	1,5	2,2	1,6	2,4	0,039
basophil leukocyte (%)	1,7	1	0,8	1,7	1,5	1,1	0,458	1	1,1	2,2	1,1	1,1	0,7	0,315

	baseline IRI + CFH dose 1			90 min IRI + CFH dose 1			p =	baseline IRI + CFH dose 2			90 min IRI + CFH dose 2			p =
Rat #	64	65	66	64	65	66		71	72	73	71	72	73	
neutrophil leukocyte (%)	9,9	6,4	7,9	23,3	18	22,8	0,003	12,2	7	16,2	28,4	24,1	40,8	0,027
lymphocyte (%)	56,4	89,7	89,1	72	77	73,6	0,724	82,1	89,3	78,8	67,2	71,5	53,5	0,036
monocyte (%)	0,9	1,2	1,4	1,5	1,83	1,2	0,214	2,3	1,7	2	0,9	1,6	2,1	0,296
eosinophil leukocyte (%)	2,7	1,7	1	2,8	0,7	1,7	0,936	2,1	1,2	2,1	2,5	2,1	2,6	0,149
basophil leukocyte (%)	0,1	1,7	0,6	0,6	1,7	0,7	0,751	1,3	0,8	1,2	1	0,9	1	0,442

Table 3 – No changes of inflammatory cell in rats injected with human CFH
EDTA-anti-coagulated whole blood of rats (3 rats per group) was drawn before (baseline) and 90 min after administration of dose 1 or dose 2 of human CFH and differential blood counting was done by an external laboratory (www.invitro.at).

6.4.2 Administered human CFH persists in rat circulation

In order to investigate why human CFH treatment did not protect from intestinal IRI, the concentration of human CFH in plasma of treated rats was determined. In rats, the pharmacokinetics of injected human CFH were not explicitly assessed, but the determination of human CFH in plasma of injected rats after 30 min, and after 90 min revealed more than 60 % and more than 50 % recovery of initally injected CFH antigen, respectively (Fig. 13).

Using this ELISA, the maximum recovery of human CFH antigen in rat circulation was reached after 30 min with an average decrease of human CFH antigen by 37 % after 90 min. This decline of human CFH was most reasonably due to renal clearance of the human protein over time.

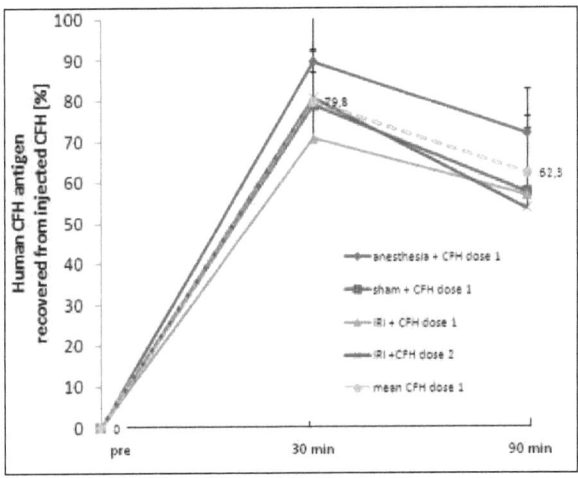

Figure 13 – Detection of human CFH in rat plasma by ELISA
Plasma of rats injected with functional human CFH drawn at onset of anesthesia before injection (pre) and after 30 and 90 min of injection was determined for human CFH antigen levels by commercial ELISA (HK342, Hycult Biotech). Data are presented as % of human CFH antigen recovered in rat plasma at indicated time points compared to injected CFH antigen (calculated) into rats and as mean values of the entire group (n=3-6 rats per group +/- SEM).

In summary, human CFH antigen could largely be recovered in plasma of all CFH-treated animals after 90 min. Nevertheless, the initial hypothesis that administration of human CFH protects rats from intestinal IRI was not proven. Differential blood counts of inflammatory cell populations of peripheral blood remained unchanged among study groups except IRI groups treated with human CFH showing an elevation of neutrophils in peripheral blood. However, no significant infiltration of inflammatory cells in the ileum or other peripheral tissues was detected by histology, suggesting no significant contribution of inflammatory blood cells to intestinal injury in this rat model.

6.5 Human CFH administered into rats subjected to intestinal IRI did not localize to sites of injury

Part of the initially hypothesized protective effect of human CFH in rat intestinal IRI implicated that administered human CFH might protect partly by targeting to sites of injury. The most vulnerable part of the gut mucosa during pathologic conditions like IRI is the gut epithelium with loss of membrane integrity, detachment of epithelial cells and subsequent loss of barrier function, finally leading to local inflammation (Collard & Gelman, 2001). Therefore, in order to exert a protective effect in this pathology, administered human CFH was assumed to bind to the intestinal epithelium.

In order to accomplish the detection of injected human CFH at tissue level, immunohistochemical localization of CFH on paraffin sections of rat intestinal tissue was established. Two commercially available antibodies served to discriminate between endogenous rat and administered human CFH in tissues. A polyclonal IgG antibody against human CFH (R&D Systems) proved to be specific for human CFH by

Western Blotting (Fig. 14, A) under both reducing and non-reducing conditions while another polyclonal IgG antibody against mouse/rat CFH (Santa Cruz) cross-reacted with CFH from rat and human sources under reducing conditions (Fig. 14, B).

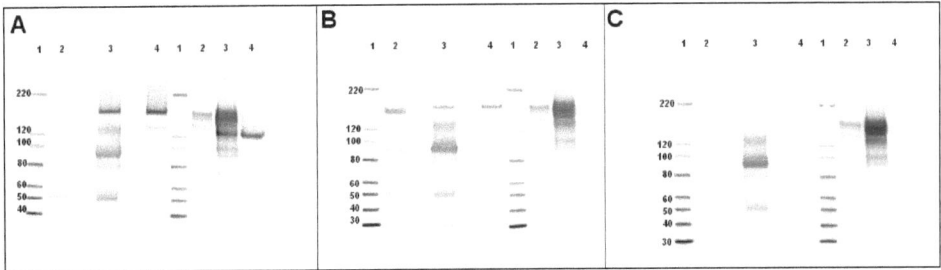

Figure 14 – Specificity of anti-CFH antibodies towards rat and human
Electrophoresis of Sprague Dawley rat plasma (lane 2), pooled human plasma (lane 3) and purified human plasma CFH (lane 4) was performed on 3-8 % TRIS-acetate gradient gel. Molecular weight markers were applied at lane 1. CFH-normalized samples were transferred onto nitrocellulose and CFH detected with (A) goat anti-human CFH IgG (R&D Systems, AF4779), (B) goat anti-mouse/rat CFH IgG (Santa Cruz, sc-17951) and (C) secondary antibody only (rabbit anti-goat IgG – HRP, Calbiochem) for 60 min. After additional 30 min incubation with secondary antibody, chemiluminescence was detected on an imaging system (Fuji, LAS-3000). The left panel of lanes 2-4 in A-C shows samples under reducing conditions. The right panel of lanes 2-4 in A-C shows samples under non-reducing conditions.

6.5.1 Detection of endogenous CFH on rat tissues

First, the ability to specifically detect endogenous CFH in rat tissue by immunohistochemistry (IHC) had to be achieved. In animals not injected with human CFH, the cross-reactive antibody stained endogenous CFH in the ileum (Fig. 15, A-C), the liver (Fig. 15, D-F), the brain (Fig. 15, G-I), the kidney (Fig. 15, J-L). Pancreatic β-islets – known to express CFH (Serrano et al, 2003)- also showed positive staining. Neither incubation with the human-specific CFH antibody (R&D Systems) nor an isotype control antibody (pre-immune goat IgG), nor incubation with a secondary antibody yielded any positive immunoreactions in these tissues. Thus, the cross-reactive antibody (Santa Cruz) specifically detected endogenous CFH in rat tissues by IHC. More specifically, endogenous CFH was detected in peri-sinusoidal capillaries of hepatocytes (Fig. 15, F), in capillary vessels of rat brain (Fig. 15, I), in vessels and partly in glomeruli of kidneys (Fig. 15, L) and in the intestinal epithelium (Fig. 15, C).

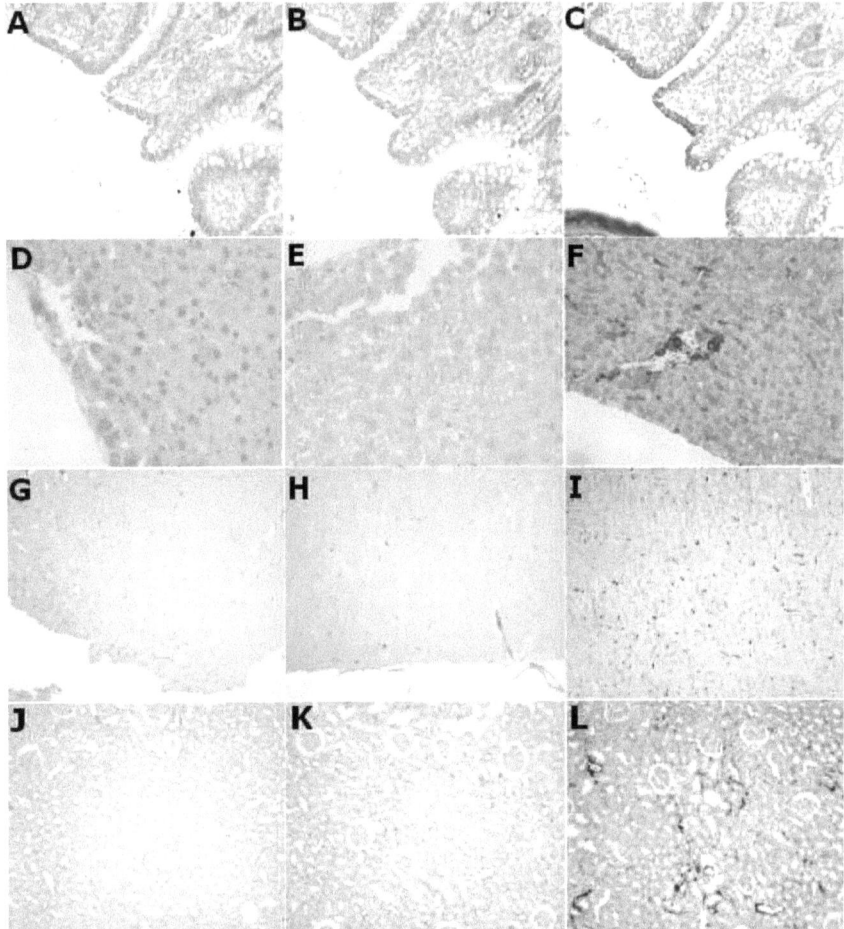

Figure 15 – Anti rat CFH antibody detects endogenous CFH in rat intestine, liver and brain

4 μm serial paraffin sections of ileum (A-C), liver (D-F), brain (G-I) and kidney (J-L) of a representative anesthetized rat without human CFH treatment were subjected to IHC with +4°C overnight incubation using (A,D,G,J) an isotype antibody (pre-immune goat IgG, Sigma, 1:200), (B,E,H,K) goat anti-human CFH (R&D Systems, AF4779, 1:200), and (C,H,I,L) goat anti-rat CFH (Santa Cruz, sc-17951, 1:200). After subsequent incubation with a biotin-conjugated secondary antibody and streptavidin-HRP, CFH was detected as brown immune-precipitate after DAB substrate exposure.

Most intriguingly, endogenous CFH was specifically immuno-localized to intestinal epithelial cells of rats subjected to anesthesia, to sham surgery and IRI surgery (Fig. 16). This staining was found to be most strongly associated with apical membranes of intestinal epithelial cells.

To the best of the author's knowledge, this is the first report of positive immunohistochemical localization of CFH in mammalian intestines. One attempt so far described to immuno-localize CFH in intestines of mice subjected to gut IRI failed due to technical difficulties (Huang et al, 2008a).

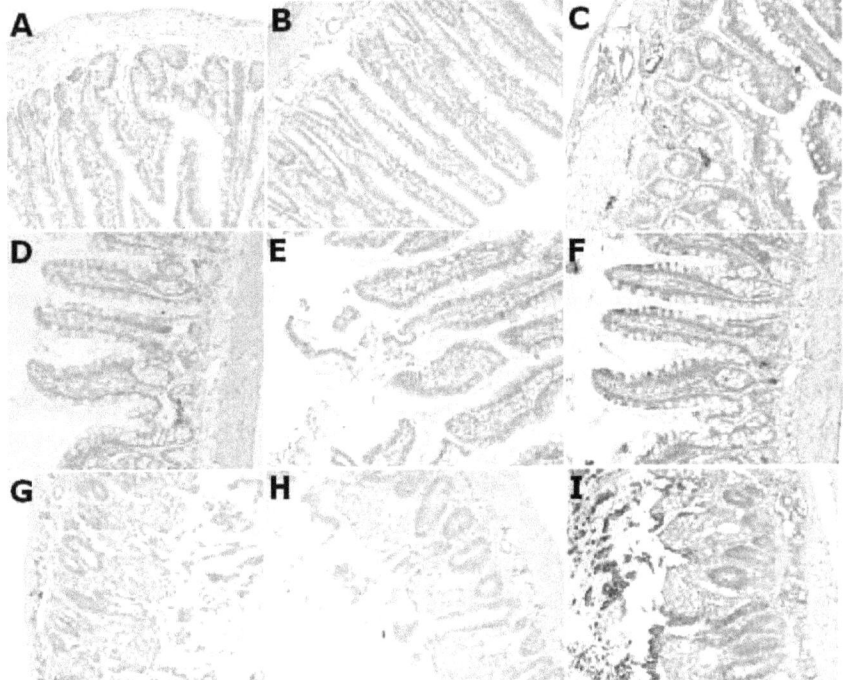

Figure 16 – Endogenous CFH along epithelia in rat ileum
Representative 4 µm serial paraffin sections of ileum of (A-C) anesthetized, (D-F) sham-operated and (G-I) IRI rats without human CFH treatment were subjected to IHC with +4°C overnight incubation by (A,D,G) isotype antibody (preimmune goat IgG, Sigma, 1:200), (B,E,H) goat anti human IgG (R&D Systems, AF4779, 1:200) and (C,H,I) goat anti rat CFH (Santa Cruz, sc-17951, 1:200). After subsequent incubation with biotin-conjugated secondary antibody and streptavidin-HRP, CFH was detected as brown immune-precipitate after DAB substrate exposure.

Finding endogenous CFH in intestinal epithelial cells implied the question whether CFH was either locally expressed by the cells – as consistently found under physiological (Fig. 16, A-C, anesthesia and Fig. 16, D-F, sham operation) and pathological (Fig. 16, G-I, IRI) conditions – or whether it adhered specifically to the gut epithelium from the circulation. Local expression of complement C3 and factor B of intestinal epithelial cells has previously been described (Andoh et al, 1998).

6.6 Human CFH did not target to intestinal sites of injury in IRI

Having identified a technical means to immuno-localize endogenous CFH in rat tissues, the author focussed on the localization of injected human CFH at the tissue level and its discrimination from rat CFH. For this reason IHC experiments using the human CFH-specific antibody (R&D Systems) on intestines of rats treated with human CFH were conducted. In Fig. 17D, a representative IHC detection of human CFH on ileum of an IRI rat treated with human CFH is shown. It can clearly be seen that human CFH – exclusively recognized by the antibody from R&D Systems (AF4779) in Western Blot (Fig. 14) – was present in the blood and lymphatic vessels along the entire mucosa. Human CFH was also found in sub-epithelial capillary and lymphatic vessels in the ileum of IRI rats that were expanded as a known result of intestinal IRI (Meng et al, 2007). Also, CFH was prominently detected on shed epithelial cells in the lumen (not visible in Fig. 17, D) albeit weaker than with the cross-reactive antibody from Santa Cruz.

Therefore, in all rats treated with human CFH irrespective of the study group, human CFH was immuno-localized exclusively to lymphatics and blood vessels detected by the human CFH-specific antibody. In contrast, in rats of all study groups not treated with human CFH, CFH was immuno-localized to lymphatics, blood vessels and in addition along the intestinal epithelium using the cross-reactive antibody. The presence of CFH along epithelia therefore could be ascribed to endogenous CFH in the rat gut, clearly discriminated from the human protein.

Administration of human CFH into rats did not result in its localization along the gut epithelium, suggesting that human CFH did not reach the sites of injury in a rat model of gut IRI. This finding provided a plausible explanation (among possibly others) why human CFH treatment did not significantly ameliorate the intestinal injury that mainly affected the gut epithelium.

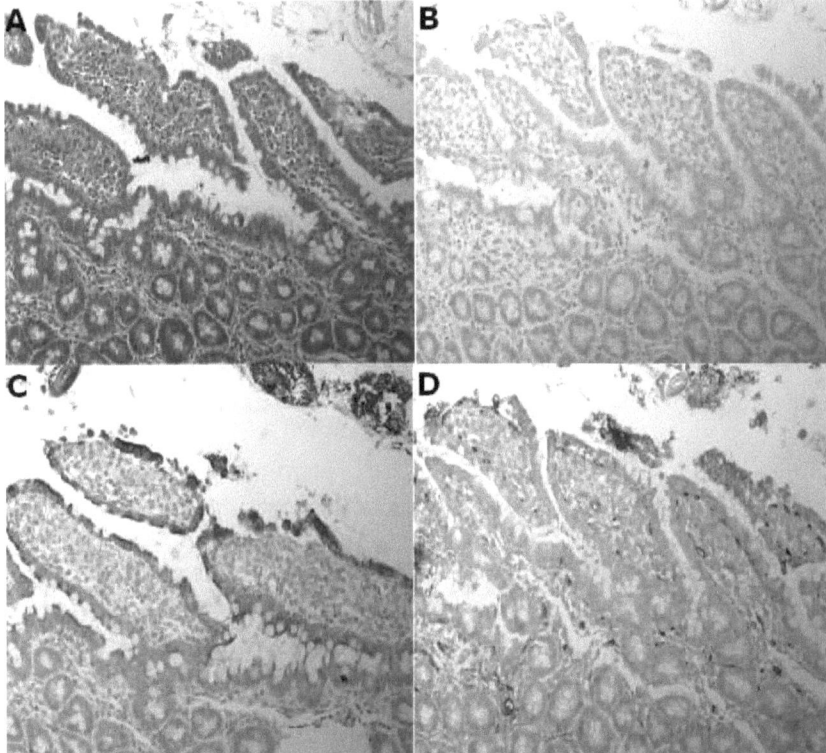

Figure 17 – Human CFH remained in intestinal blood vessels and did not target the epithelium in gut IRI Representative 4 μm serial paraffin sections of ileum of an IRI rat treated with human CFH dose 1 were H+E stained (A) or subjected to IHC with +4°C overnight incubation by (B) isotype antibody (pre-immune goat IgG, Sigma, 1:200), (C) goat anti rat IgG (Santa Cruz, sc-17951, 1:200) and (E) goat anti human CFH (R&D, AF4779, 1:200). After subsequent incubation with biotin-conjugated secondary antibody and streptavidin-HRP, CFH was detected as brown immune-precipitate after DAB substrate exposure.

6.7 Human CFH did not target renal sites of injury in IRI

In addition to the intestinal disease model, CFH treatment was performed in an established rat model of renal IRI in the laboratories of a clinical research organization in Taiwan (Eurofins PanLabs formerly Ricerca Biosciences LLC, Bothell, WA 98021, USA, Pharmatoxicology division). Previously, it has been found that complement activation after renal IRI is exclusively caused by the alternative complement pathway (Thurman et al, 2003a).

The rationale for human CFH treatment of rats subjected to renal IRI is based on previous studies of CFH-based treatments of renal IRI in mice, which demonstrated an amelioration of the underlying kidney injury (Renner et al, 2011).

The rat model of renal IRI was conducted as described (Chintala et al, 1994; Nakamoto et al, 1987; Tripatara et al, 2007). Intravenous administration of purified human CFH at dose 1 (700 µg human CFH per ml rat plasma) did neither significantly change urine levels of kidney injury molecule-1 (KIM-1) and N-acetyl-β-D-glucosaminidase (NAG), two clinically relevant sensitive biomarkers for acute renal failure (Han et al, 2002), nor ameliorate blood urea nitrogen (BUN) and creatinine levels in serum.

Normal architecture of kidneys was detected in sham controls, whereas IRI rats revealed massive to severe necrosis affecting mainly the proximal tubules, severe high hyaline casts and tubular mineralizations. However, no significant differences in renal histopathology were detected between CFH-treated and vehicle-treated IRI rats by two-tailed Student´s T-test ($\alpha = 0.05$).

Thus, the localization of human CFH in kidneys of treated rats was investigated by IHC. In line with the absence of human CFH at sites of intestinal injury (Fig. 17), no immunohistochemical localization of human CFH at the disrupted tubular epithelium as major site of injury in the kidney model was observed (Fig. 18, arrows points at tubular epithelial cells).

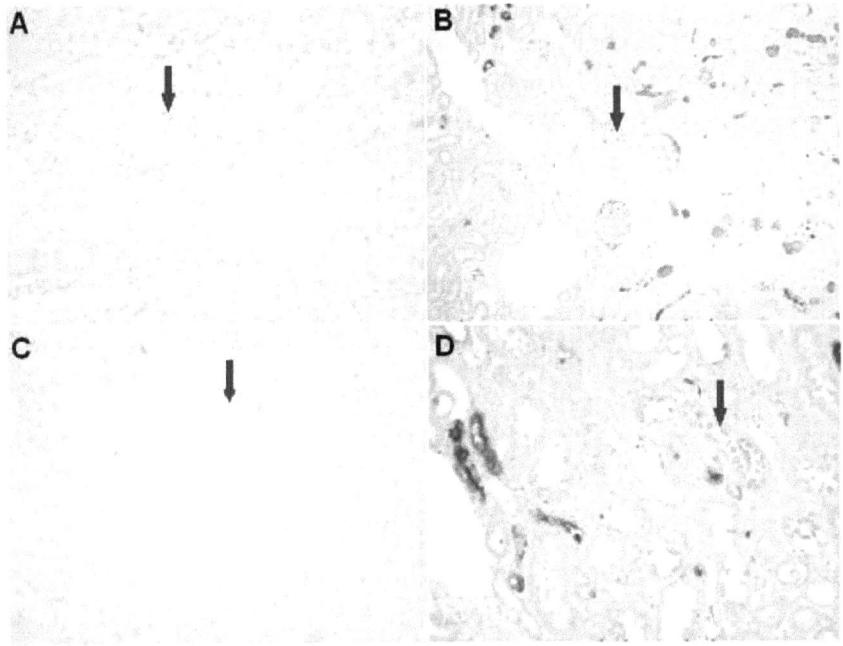

Figure 18 – Human CFH did not target renal sites of injury after renal IRI
Representative 4 µm serial paraffin sections of kidney of a rat subjected to renal IRI and treated with human CFH dose 1 (700 µg/ml) were incubated with (A) an isotype antibody (pre-immune goat IgG, Sigma), (B) goat anti human CFH (R&D Systems, AF4779), (C) antibody dilution buffer only and (D) goat anti rat CFH (Santa Cruz, sc-17951). After subsequent incubation with a biotin-conjugated secondary antibody and streptavidin-HRP, CFH was detected as brown immunoprecipitate after DAB substrate exposure. Arrow points at disrupted tubular epithelial cells.

In summary, an immunohistochemical discrimination of endogenous rat from exogenous human CFH in rat tissues was established.

It was consistently shown that human CFH injected into rats remained in circulation and did not target the intestinal epithelium where endogenous CFH could be found either as result of local expression or adhesion from circulation. Likewise, injected human CFH did not reach renal sites of injury in an established rat model of renal IRI.

In conclusion, one reason why human CFH administration did not protect from respective clinical outcomes in relevant rat models of IRI may be the fact that the human CFH did not reach the sites of injury in these models.

6.8 Caco-2 cell model of human intestinal epithelial cells expresses CFH

The novel finding of CFH immuno-localization along gut epithelia, suggested the possibility that CFH is made by epithelial cells. Therefore, CFH expression in a human intestinal epithelial cell was determined. Caco-2 cells are human colonic carcinoma cells that upon maturation largely resemble human intestinal epithelial cells (Artursson et al, 2012) and were shown to express complement proteins C3 and factor B (Andoh et al, 1998; Moon et al, 1997). Therefore, Caco-2 whole cell lysates (Abcam, ab-3950) were separated by SDS-PAGE and CFH expression determined by Western blotting using a polyclonal antiserum to human CFH (Comp Tech, A337). Fig. 19 shows that Caco-2 cells express CFH protein. These results corroborated the novel finding of CFH immuno-localization along intestinal epithelia in rats (Fig. 16), which suggested local CFH expression by intestinal epithelial cells.

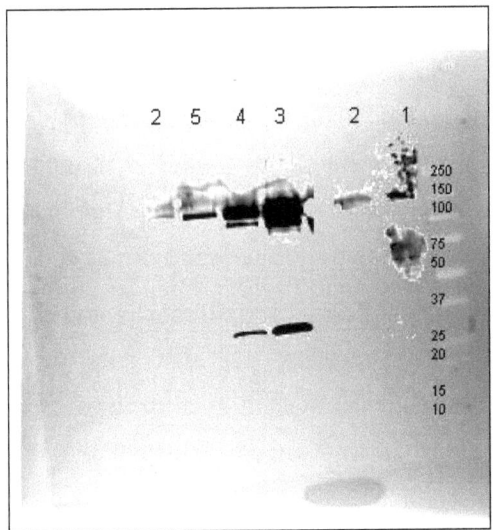

Figure 19 - Caco-2 cell model of human intestinal epithelial cells expresses CFH

Caco-2 whole cell lysates (Abcam, ab3950) in lanes 1 (20 µg total cellular protein) and 2 (10 µg total cellular protein) and purified human CFH (lane 3, 100 ng; lane 4, 10 ng; lane 5, 1 ng) were separated on 4-20 % TRIS-glycine gel (Invitrogen), transferred on nitrocellulose and probed with goat antiserum to human CFH (Comp Tech, A337) for 60 min. Specificity control blot by probing with secondary antibody showed no chemiluminescence.

6.9 Human CFH bound dying intestinal epithelial cells in intestinal IRI

Binding of CFH to apoptotic (Leffler et al, 2010; Martin et al, 2012; Mihlan et al, 2009; Trouw et al, 2007) and necrotic cells (Lauer et al, 2011; Weismann et al, 2011) has been demonstrated before. Therefore, binding of CFH to dying cells of the ischemic intestine was investigated.

Serial paraffin sections from the ileum of IRI rats and IRI rats treated with human CFH were prepared. Dying intestinal epithelial cells were identified on hematoxylin and eosin (H+E) stained sections based on their typical morphology. Dying epithelial cells were detected as shed cells in the lumen, partly displaying already condensed, pyknotic or fragmented nuclei (Fig. 20, C). Subsequently, serial sections of these tissues were immunostained for CFH using the antibody specific for human CFH from R&D Systems (Fig. 20, A) and the antibody cross-reactive between human and rat CFH (Fig. 20, B). Note that in all groups of rats treated with human CFH, human CFH could only be detected in blood and lymphatic vessels.

Human CFH did not target intestinal epithelial cells as major site of injury of intestinal IRI (Fig. 17). However, human CFH did bind dying epithelial cells in the ischemic ileum (Fig. 20, A). This is consistent with reports, demonstrating the presence of CFH on cellular debris of necrotic areas in human adeno-carcinomas of the colon (Wilczek et al, 2008).

Rat CFH was identified to bind to dying intestinal epithelial cells as exemplified in Fig. 20, B. Interestingly, despite the fact that injected human CFH did not bind to healthy epithelial cells under physiological or pathological conditions, it obviously achieved binding to dying epithelial cells of ischemic rat intestines. Despite the absent protection from intestinal injury by human CFH treatment, these findings indicate the retained functionality of administered CFH.

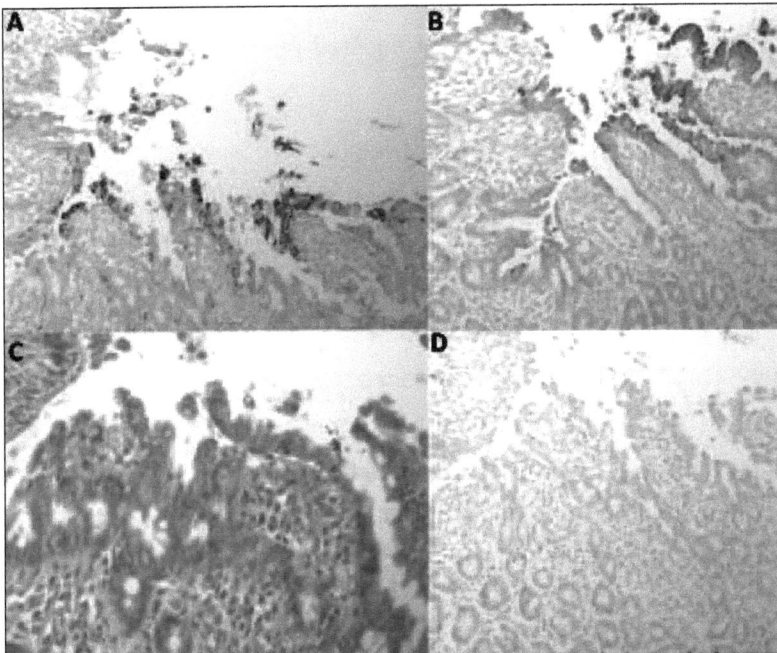

Figure 20 – Rat and Human CFH focally bound dying intestinal epithelial cells in intestinal IRI

(A) Representative IHC staining of human CFH in the ileum of a human CFH treated IRI rat using a human CFH-specific antibody (R&D Systems), 10x magnification.

(B) Representative IHC staining of CFH in the ileum of the same rat using a rat/human-crossreactive antibody (Santa Cruz), 10x magnification

(C) Representative H+E staining of a serial section of the ileum of the same rat, 20x magnification

(D) Representative IHC staining of another serial section of ileum of that rat using isotype control antibody (goat IgG), 10x. Presented pictures A-D are consecutive serial sections (4 μm) from one representative IRI rat treated with human CFH.

6.10 Injected human CFH into rats subjected to intestinal IRI maintained functionality in rat circulation *in vivo*

Having clarified the effects of human CFH injection in intestinal IRI at the tissue level, the functional ability of injected human CFH in regulating complement was evaluated.

For this reason, an assay was developed to facilitate the precise determination of human CFH activity in rat plasma. Interaction of human CFH with rat complement has been shown before (Granoff et al, 2009). The assay based on the assumption that rat plasma by itself – although containing endogenous CFH at human-like concentrations (Daha & van Es, 1982; Demberg et al, 2002) – induces strong lysis of sheep red blood cells (SRBC). Spiking of purified human CFH reverses rat plasma-mediated SRBC hemolysis dose-dependently and therefore demonstrates the ability of human CFH to regulate rat complement. The same purified CFH as administered into rats was used to generate a standard curve in this assay. In Fig. 21, the mean hemolysis curve of three standard preparations of purified human CFH spiked into rat plasma is presented as linear regression of absorbance and known concentrations of spiked human CFH. The inset of Fig. 21 illustrates the mean reduction of human CFH activity in plasma obtained from IRI rats treated with human CFH at dose 1 (700 µg human CFH per ml rat plasma) after 90 min. This reduction of human CFH activity between time points was found insignificant when compared with the standard curve ($p = 0.055$, two-tailed T-test) and was found to be 33 % (n = 7 rats). Likewise, the mean reduction of human CFH activity in plasma of IRI rats treated with human CFH at dose 2 (350 µg human CFH per ml rat plasma) was 30 % (n=3 rats).

For comparison, the mean reduction of human CFH activity in plasma of anesthetized and sham-operated rats treated with dose 1 human CFH was 43 % and 45 %, respectively (n = 3 rats each) and moreover found to be significant when compared with the standard curve (sham + CFH vs. standard, p = 0.001; anesthesia + CFH vs. standard, p = 0.013).

The decrease of human CFH in rat plasma after 90 min is consistent with its renal clearance. Indeed, strong immunohistochemical localization of human CFH in renal glomeruli of human CFH-treated rats was observed.

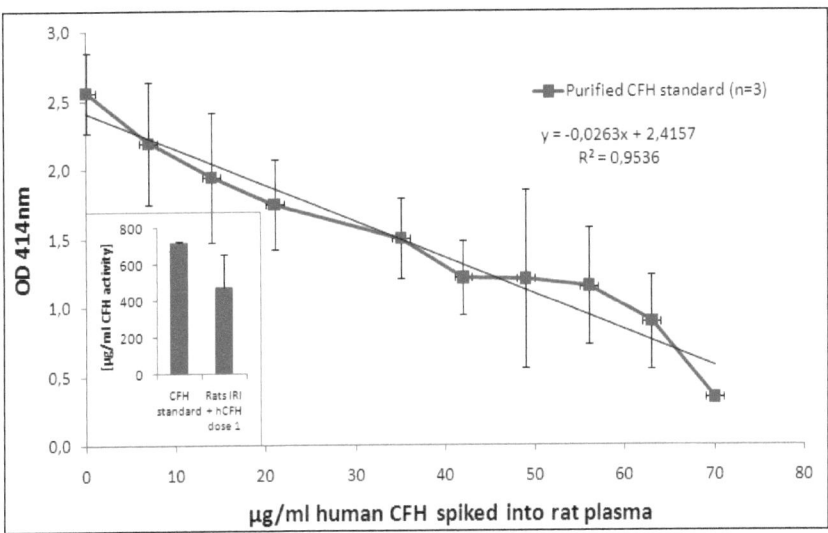

Figure 21 – Human CFH maintained hemoprotective activity after 90 min within plasma of IRI rats Plasma of IRI rats injected with 700 µg human CFH per ml rat plasma (dose 1) or 350 µg/ml as dose 2 obtained after 90 min (30 min ischemia + 60 min reperfusion) was diluted with preimmune rat plasma and incubated with SRBC in PBS containing 7 mM $MgCl_2$ and 10 mM EGTA for a total of 60 min at +37°C. To generate a standard curve, purified CFH was spiked into pre-immune rat plasma and diluted in the same way as rat plasma samples containing human CFH. Complement-mediated hemolysis of SRBC was stopped by adding 20 mM EDTA. Hemolysis of supernatants was determined as absorbance at 414 nm. Inset shows the average CFH activity in plasma of 7 IRI rats treated with human CFH at dose 1 compared to the average of three standards curves.

The ability of administered human CFH to regulate complement in rat plasma obtained 90 min after intestinal IRI demonstrated that administered human CFH largely maintained its activity during 90 min as assessed by an *ex vivo* assay.

6.11 Human CFH administration into rats subjected to intestinal IRI prevented local complement deposition at ischemic intestine

Strong local complement deposition in ileum of IRI rats has been demonstrated (Fig. 11). Although injected human CFH did not protect from mucosal injury (Fig. 12), it was shown to largely maintain biological activity in rat circulation *ex vivo* (Fig. 21).

In light of this, it was tempting to investigate whether injected human CFH provided protection from local complement deposition at the ischemic rat intestine. Therefore, C3 deposition was assessed on frozen sections of rat ileum using a FITC-conjugated polyclonal antibody to human C3c (Abcam, ab4212) cross-reactive to rat C3. Fig. 22, C and D demonstrate that human CFH treatment indeed effectively reduced C3 activation in rat ileum as seen of rats undergoing IRI (Fig. 22, A and B).

The strong C3 staining along the villi tips of IRI rats (compare Fig. 11, C and D and Fig. 22, A and B) including sub-epithelial capillaries, dilated vessels and epithelial cells was not observed in the villi of IRI rats treated with human CFH. Anesthetized and sham-operated rats treated with human CFH also showed very slight C3 staining in the vessels (Fig. 11, A and B) and were comparable to CFH-treated IRI rats.

These results provided further evidence for the activity of injected human CFH *in vivo*. Moreover, the data suggest that local complement deposition as result of intestinal IRI in rats did not contribute significantly to the underlying tissue pathology.

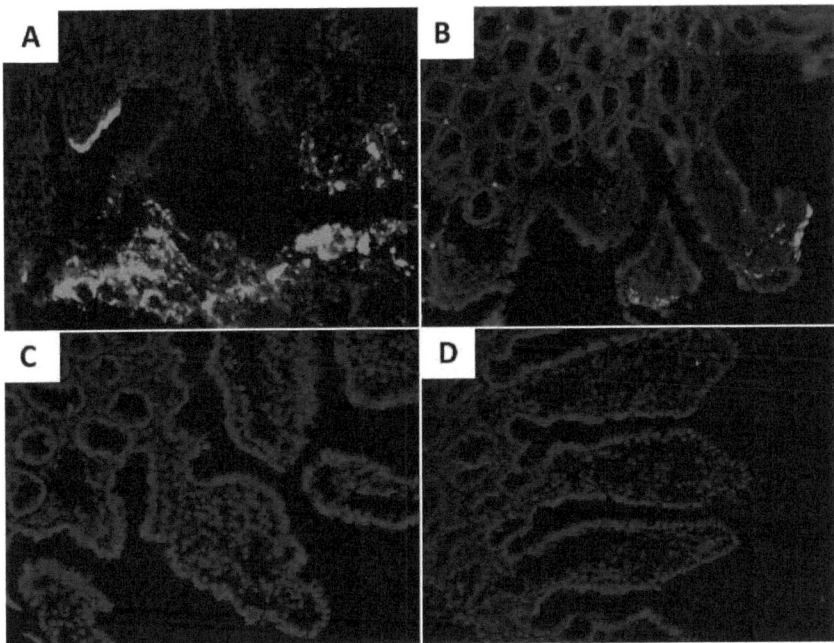

Figure 22 – Human CFH protected from local complement deposition on ischemic rat ileum

Representative immunefluorescence microscopic photographs (20x magnification) of C3 deposition in ileum of two IRI rats (A, B) and of two IRI rats injected with human CFH (C, D). Immunefluorescence experiment was conducted as described under methods and was representative of 3 independent experiments. Equal microscopic magnifications on all slides were applied (20x).

7 DISCUSSION

The role of complement in the pathology of mesenteric IRI has been extensively demonstrated in the last decade in mice (Austen et al, 2003; Lee et al, 2010; Lu et al, 2008), rats (Eror et al, 1999; Hill et al, 1992; Padilla et al, 2007; Souza et al, 2005; Wada et al, 2001) and humans (Grootjans et al, 2010; Matthijsen et al, 2009). A CFH-based complement inhibitor has been shown to protect mice from intestinal IRI, though the exact mechanisms of protection were not elucidated (Huang et al, 2008a). In this work, treatment of a rat model of gut IRI with highly pure human CFH was performed to investigate the effect of human CFH in IRI protection and complement control.

NOVEL PREPARATION OF NATIVE CFH

In order to supply highly purified CFH for the preclinical studies, a novel and scalable CFH preparation process was developed that resulted in purification of fully functional CFH. Of note, native CFH was purified from side fractions of industrial plasma fractionation, which would enable large scale production. Efficient pathogen safety steps, S/D treatment and nanofiltration, were included in the purification process that did not ameliorate CFH activity.

The process moreover achieved the specific removal of dysfunctional CFH species by a certain chromatographic step. To the knowledge of the author, the significant reduction of the truncated and dysfunctional CFH species has not been shown before with any other CFH purification process. The CFH concentrate was found to be fully active in terms of cofactor activity mediating CFI-dependent C3b cleavage, acceleration activity of C3 convertases decay and SRBC protection from complement-mediated in vitro hemolysis. In addition, the CFH preparation was confirmed to be free from endotoxin and thus was suited for the in vivo study in rats and in vitro studies involving cell cultures.

RAT MODEL OF COMPLETE AND WARM ISCHEMIA REPERFUSION IN GUT

The newly developed CFH preparation was studied in a rat model of intestinal IRI. In this model, complete and warm ischemia was created by filament ligation of collateral blood supply together with restriction of the arterial supply by a microvascular clamp. Tight closure of the abdomen after ischemia as well as after reperfusion phase ensured warm ischemia. Previous studies delineated the high level of destructiveness of warm IRI by the fact that severe mucosal injury in rats was achieved already after 30 min of total warm ischemia followed by 60 min of reperfusion (Beuk et al, 2000) while 24 hours of cold ischemia were needed to induce the respective extent of tissue injury in rat ileum (Massberg et al, 1998).

The importance of restricting also collateral blood supply to create complete ischemia in these models was highlighted by others referring to the non-feasibility of severe mucosal lesions by ligation of the superior mesenteric artery (SMA) only (Deitch et al, 1987; Yamamoto et al, 1980) because of collateral supply (Leite Junior et al, 2010). Insufficient restriction of collateral supply plausibly caused lower intestinal injury as observed in two IRI rats in this study. Similar rat models of gut IRI focusing on complete occlusion of the terminal ileum as developed in this work have been described (Boros et al, 1995; Leite Junior et al, 2010). For comparison of intestinal injury of the same topographical segment of the small intestine of animals of all study groups, IRI was created in the terminal ileum.

In *vivo* imaging of reversible and irreversible cellular responses after complete IR in the murine jejunum provided evidence for the heterogeneity of vulnerability along the villus-crypt axis *in vivo* (Guan et al, 2009). This possibly explains part of the heterogeneity of mucosal injuries induced in similar gut IRI models reported in the literature. In order to avoid possible sources for artificial heterogeneity of tissue injury and lesion enlargement *post mortem*, intestinal tissue handling was performed at the utmost reproducible manner including time of tissue dissection, size of tissue samples and tissue fixation procedures (Hewitt et al, 2008; Leite Junior et al, 2010; Werner et al, 2000).

GRADING OF INTESTINAL INJURY / INFLAMMATORY CELLS

The mucosal tissue injury was graded by a well accepted scoring system (Chiu et al, 1970). The surgical protocol induced significantly higher intestinal injury scores in IRI rats compared to sham controls and hence allowed the identification of possible differences of more than one Chiu score unit between treated and untreated groups.

This mucosa damage of IRI rats was induced locally in ileum. No damage was found in internal control tissue of duodenum and jejunum. Likewise, pathologic evaluation of peripheral organs revealed no sign of remote organ injury. The absence of any remote organ damage in this rat model, however, did not reflect the prevailing clinical presentation of human intestinal IRI in which reperfusion injury often elicits a systemic inflammatory response syndrome that mostly progresses to multiple organ failure (Stallion et al, 2005).

No relevant changes in leukocytic cell populations in peripheral blood were detected among human CFH treated groups except a significant increase of neutrophils in blood of human CFH-treated IRI rats after 90 min. However, the variance of cell numbers between these groups at both time points argues against significance. Likewise, a histological evaluation revealed no significant elevation of inflammatory cells in tissues after the 90 min procedure in any study group. These findings indicated that inflammatory cells were not recruited to intestinal sites of injury as shown in previous reports and hence do not contribute to the pathology in this model.

NO SYSTEMIC COMPLEMENT ACTIVATION BUT LOCAL COMPLEMENT DEPOSITION IN RATS

A moderate decline of complement hemolytic activity (CH50) of about 10 % in rats subjected to intestinal IRI was recently reported and a CH50 decline of above 10 % in experimental models was suggested to already indicate systemic complement activation (Ehrnthaller et al, 2012). In this study, the reductions in CH50, C3 concentrations and total protein levels in plasma after 90 min were comparable in all study groups, were moreover reflected by plasma expansion in human CFH treated groups, and were therefore not indicative of systemic complement activation. This was sustained by the fact that administered human CFH, although potently regulating rat complement in circulation, did not affect these reductions of CH50 and plasma C3 concentrations in treated IRI rats. Seemingly, there was no need for systemic complement regulation in this rat model.

In contrast, a strong local complement deposition specifically at ischemic ileum as result of IRI surgery was elicited in agreement with previous studies (Atkinson et al, 2005a; Huang et al, 2008a; Souza et al, 2005). The area of C3 deposition within the ischemic bowel along epithelial cells of villi tips was in accordance with previous reports (Fleming et al, 2010; Pope et al, 2012).

RATIONALE FOR CFH THERAPY IN THE INTESTINAL MODEL

The implication of marked mucosal injury and local complement deposition in this rat model provided a rationale for testing CFH as therapeutic complement inhibitor. This rationale was in part based on the hypothesis that local complement activation in the rat intestine may be causative for and/or strongly contributing to mucosal tissue injury (Gorsuch et al, 2012).

In support of this, complement intervention at the C3 level (Atkinson et al, 2005a; Huang et al, 2008a; Stahl et al, 2003) and C5 level (Wada et al, 2001) has been proven to successfully ameliorate tissue injury in gut IRI models in rodents. A previous report claimed that endogenous CFH failed to protect from intestinal IRI (Huang et al, 2008a). The authors of that report could show protection from intestinal injury by a "targeted" CFH-based complement inhibitor, but did not provide an exact mechanism of action, nor data on the activity of this inhibitor *in vivo*. The present report aimed to overcome these limitations.

In addition, the rationale for CFH therapy in the intestinal model is also based on the fact that IRI mediates acceleration of the restitution process of intestinal epithelial cells (Iizuka & Konno, 2011) undergoing shedding (Matthijsen et al, 2009) and mainly apoptosis (Grossmann et al, 2002; Ikeda et al, 1998). This accelerated process leads to an extended gradient of shed epithelial cells into the gut lumen and possibly results in a loss of endogenous CFH following this gradient. Direct assessment of this hypothesis would have required the tracking of e.g. fluorescently labeled CFH by intra-vital fluorescence microscopy (Kalia et al, 2002; Szabo et al, 2008) over time *in vivo*.

These hypotheses would support the notion that higher doses of an already abundant plasma protein like CFH should be administered in intestinal IRI for therapeutic purposes. Two doses of human CFH were applied to investigate a dose-dependent protective effect of CFH on mucosal injury. These dosages were extrapolated from mouse models of human disease injected with purified human CFH (Fakhouri et al, 2010b; Griffiths et al, 2009).

INTRAVENOUS HUMAN CFH ADMINISTRATION DID NOT RESCUE RATS FROM GUT IRI

However, the hypothesis of protection by human CFH therapy of intestinal IRI in rats was not confirmed, as no significant protection by administration of human CFH at therapeutic doses was observed. Subsequent studies attempted to provide explanations for this finding.

HUMAN CFH REGULATED RAT COMPLEMENT IN FLUID PHASE

Loss of or diminished activity of human CFH in the rat circulation would have provided one possible explanation. However, the recovery of human CFH antigen in plasma obtained from human CFH-treated rats was documented. Moreover, human CFH activity was also found to be maintained to large extent in rat plasma. On average 70 % of initially injected human CFH antigen and of human CFH activity was detected after 90 min in rat circulation. The reduction of the human protein in rat plasma was likely due to renal clearance. Indeed, marked staining of human CFH in renal glomeruli of human CFH-treated rats of all groups was detected by IHC.

The hemolysis assay developed to determine human CFH activity in rat plasma had two benefits. First, it demonstrated efficient regulation of rat complement by human CFH *ex vivo*. Second, the assay allowed the selective detection of human CFH activity, as rat plasma *per se* fully lysed SRBC in solution. Therefore, accurate determination of active human CFH in rat plasma could be achieved. Western blotting revealed no CFH-specific band except native CFH in plasma of CFH-treated rats indicating its molecular integrity (not shown).

In summary, it was proven that injected human CFH largely maintained function in rat circulation after 90 min. Surprisingly, given the absence of protection from gut injury, human CFH administration of dose 1 (700 μg human CFH per ml rat plasma) was able to completely abrogate local complement deposition at ischemic ileum.

It was therefore concluded that local complement deposition was a specific result of ischemia reperfusion at the distal ileum but did not have significant effect on IRI-induced mucosal injury in this rat model. In fact, the local complement deposition observed at this static and early time point after reperfusion can be regarded as complement-mediated opsonization of altered self following oxidative stress in order to favor its clearance by phagocytes.

HUMAN CFH DID NOT TARGET INTESTINAL SITES OF INJURY

In the course of finding explanations for the lack of protection of human CFH treatment in this model, the localization of human CFH at tissue level was investigated. Using IHC, it was found that human CFH exclusively localized to either lymphatic and blood vessels and sub-epithelial capillaries along the entire gut mucosa, but not to the intestinal epithelium as most relevant site of tissue damage after intestinal IRI. Therefore, the main reason for the lack of protection by CFH therapy in this rat model is believed to be its inability to target the intestinal epithelium.

HUMAN CFH DID NOT TARGET RENAL SITES OF INJURY

Additional evidence that injected human CFH did not target sites of injury in disease models came from a subsequent study testing CFH in an established and clinically relevant rat model of renal IRI (Chintala et al,

1994; Nakamoto et al, 1987; Tripatara et al, 2007). Therein, injected human CFH did also not localize to deteriorated tubular epithelial cells in rat kidneys as major site of renal injury nor did CFH treatment improve several sensitive markers of acute kidney injury in the urine and blood.

In the kidney model reperfusion was performed for 2 days, and determinations of human and rat CFH by IHC were performed at this single time point only. Therefore, one could have missed to observe any protective effect by CFH on the tissue level at earlier time points. However, unchanged urine and blood markers of renal injury during the entire study period in CFH-treated IRI rats when compared to vehicle-treated IRI controls argue against this notion.

ENDOGENOUS CFH LOCALIZED TO INTESTINAL EPITHELIUM

In contrast to the injected human CFH, a specific localization of endogenous CFH on intestinal epithelial cells was detected by IHC. Of note, a previous study applying a recombinant CFH-based inhibitor in a mouse model of gut IRI failed to detect endogenous CFH in intestine for technical reasons (Huang et al, 2008a). Two commercial antibodies enabled discrimination between endogenous CFH from human CFH by IHC. Both antibodies displayed CFH staining in blood and lymphatic vessels. Although cross-reactive in Western blotting for rat and human CFH sources, only the rat-specific CFH antibody was able to specifically immune-localize CFH in various tissues of known CFH expression. The epithelial CFH staining prompted the investigation of CFH expression by intestinal epithelial cells. Weak CFH expression of Caco-2 cells, a well recognized model cell line for human small intestinal epithelial cells (Artursson et al, 2012) shown to express complement proteins C3 and factor B (Andoh et al, 1998; Moon et al, 1997), was detected by Western blot.

It would be of great interest to investigate whether primary small intestinal epithelial cells are indeed capable of expressing CFH and to understand the possible function of intestinal CFH in this regard.

HUMAN CFH BOUND DYING INTESTINAL EPITHELIAL CELLS

Administered human CFH was not found to bind to intestinal epithelium. However, binding to dying intestinal epithelial cells in ischemic intestines of CFH-treated rats could be detected by IHC. Notably, this binding was found in epithelial cells with histological signs of cell death like fragmented nuclei. In agreement, prominent binding of CFH to apoptotic (Leffler et al, 2010; Martin et al, 2012; Mihlan et al, 2009; Trouw et al, 2007) and necrotic cells (Lauer et al, 2011; Weismann et al, 2011) has been shown before. This finding provided further evidence for retained functionality of administered human CFH.

SUMMARY AND CONCLUSIONS

In summary, a novel method to isolate CFH from human plasma fractions was developed that uniquely achieved depletion of dysfunctional CFH species and purification of largely native CFH. The CFH preparation was applied as therapeutic complement inhibitor in a rat model of intestinal IRI with significant local mucosal injury and local complement deposition in the ileum but without remote organ injury.

The initial hypothesis that intravenous administration of human CFH would protect rats from intestinal IRI was not confirmed. Moreover, the intestinal injury was not found to result from local complement deposition as injected human CFH abrogated that deposition. Although not protecting from intestinal injury, human CFH maintained its integrity and ability to regulate rat complement in fluid phase. Focal binding of dying intestinal epithelium in IRI by human CFH also provided evidence for retained functionality of human CFH *in vivo*. The rat model did not elicit systemic complement activation, as no difference in complement activity between IRI rats treated or not with human CFH was found.

An explanation as to why human CFH treatment did not rescue rats from intestinal IRI was provided by the finding that human CFH did not achieve to reach the intestinal epithelium as major site of injury in this pathology. Likewise, injected human CFH did not target to renal sites of injury in a rat model of ischemic kidney injury.

Thus, it should be concluded that CFH therapy appears more reasonable in models of uncontrolled complement activation in the circulation demanding fluid phase regulation, such as in paroxysmal nocturnal hemoglobinuria (Risitano et al, 2012), atypical hemolytic uremic syndrome (Heinen et al, 2012; Hirt-Minkowski et al, 2010) or CFH deficiency (Fakhouri et al, 2010b).

8 FUTURE DIRECTIONS

In this section the author provides an outlook on aspects of future research in part based on and related to findings of this work.

IS CFH LOCALLY PRODUCED BY INTESTINAL EPITHELIUM?

Related to the novel finding that endogenous CFH could be detected along the apical gut epithelia, further investigations should address the possibility of local CFH expression by intestinal epithelium directly. This appears likely provided by the fact that human monocytes (Whaley, 1980), keratinocytes (Timar et al, 2006), fibroblasts (Katz & Strunk, 1988), endothelial cells (Brooimans et al, 1990) and retinal pigmented epithelial (RPE) cells (Chen et al, 2007; Kim et al, 2009; Mandal & Ayyagari, 2006) are all sources of extra-hepatic CFH expression.

Future research should address whether primary small intestinal epithelial cells are indeed capable of expressing CFH and moreover to elucidate the potential function of CFH at this site. Provided that intact gut epithelia are the most important component of sustained mucosal barrier function (Turner, 2009), one might speculate that CFH participates in the regulation of the apical junctional complex's function to maintain intact tight junctions. It would be interesting in this respect to assess whether CFH-deficient mice develop deteriorated barrier function over time.

ROLE OF CFH FOR INTESTINAL PERFUSION?

Additional future research should explore the reasons why human CFH remained only in circulation especially in the villous microcirculation like sub-capillary vessels. This may be due to the fact that intestinal IRI manifests very early in microvascular defects affecting the mucosal villous microcirculation (Beuk et al, 1997; Beuk et al, 2000).

In this view, the predominant localization of human CFH in intestinal microcirculation, might point to a role of CFH for intestinal perfusion as has been shown for the retina (Lundh von Leithner et al, 2009). There, CFH-deficient mice exhibited an accumulation of C3 and C3b on the vascular endothelium of the neural retina over time in combination with a constriction of retinal capillaries and venules thereby reducing retinal perfusion. Thus, it would be interesting to test the effects of CFH on intestinal microvasculature using CFH-deficient mice.

WHAT EXACTLY TRIGGERS COMPLEMENT DEPOSITION ON ISCHEMIC INTESTINE?

Crucial aspects of future research related to this work will deal with the fundamental importance of complement in the pathogenesis of IRI as such and of intestinal IRI in particular.

There is some degree of controversy as to which molecules and complement pathways are involved in the initiation of complement activation following ischemia reperfusion in the literature (Gorsuch et al, 2012). In this study, ischemia reperfusion of the terminal ileum induced local complement deposition along vulnerable mucosal structures comprising epithelial cells at the villous tip. However, neither the identity of these C3-positive structures was investigated nor what exactly elicits local complement deposition on ischemic intestine. Local complement deposition was only determined at the final endpoint after 90 min reperfusion and could therefore simply reflect normal complement function of opsonizing altered self for removal.

A thorough research on what exactly triggers complement deposition and/or activation in the reperfused intestine after ischemia has to investigate the structures that are recognized and the time point of complement deposition during intestinal ischemia reperfusion.

Does down-regulation of endogenous complement regulators under oxidative stress trigger complement deposition on ischemic intestine?

Down-regulation of membrane-bound complement regulators on human endothelial cells (Thurman et al, 2009; Vakeva & Meri, 1998), retinal pigmented epithelial cells and renal tubular epithelial cells (Renner et al, 2010) under oxidative stress rendered these cells susceptible to complement-mediated attack and has been shown to be a potential initiator of complement activation in conditions of IRI.

In this study, down-regulation of endogenous regulators of the intestinal epithelium as potential trigger for local complement deposition was not investigated. However, it will be worthwhile to do so in future research assuming that a significant down-regulation of endogenous complement regulators might indeed provide a potential trigger for complement activation in intestinal IRI.

Do Malondialdehyde epitopes trigger complement activation at ischemic intestine?

Recently, it has been found that CFH binds to the prominent lipid peroxidation product malondialdehyde (MDA) and protects from MDA-associated pro-inflammatory effects *in vitro* and *in vivo* (Weismann et al, 2011). CFH strongly bound MDA epitopes present on necrotic and apoptotic cells or present in lesions of AMD patients, which suggested that MDA mediates recognition of dying cells by CFH (Weismann et al, 2011).

It would be highly interesting to investigate expression of MDA epitopes along intestinal tissue prone to intestinal IRI and whether CFH binds MDA-decorated intestinal tissue thereby preventing MDA-mediated complement activation. It has been shown that CFH inactivated complement on MDA-decorated surfaces like on apoptotic cells (Weismann et al, 2011).

It can be speculated that dying epithelial cells of IRI rats shown here to bind human CFH were covered by MDA epitopes. Formation of MDA epitopes due to intestinal IRI appears very likely. Elevated MDA levels of intestinal, liver and lung tissue after intestinal IRI have been reported (Boyuk et al, 2011; Horton & Walker, 1993; Sener et al, 2001; Tekin et al, 2009). A recent lipidomics study revealed major changes in intestinal lipid composition of jejunum in mice subjected to intestinal IRI compared to sham controls (Sparkes et al, 2010) . Oxidation of low density lipoproteins (oxLDL)– with MDA being a major oxidation-specific epitope on LDL (Palinski et al, 1990) – and deposition of oxLDL on distal ileum, lungs and even liver of rats subjected to intestinal IRI (Tekin et al, 2009) is also supportive of the notion that enhanced lipid peroxidation might be a crucial link to inflammation after mesenteric IRI.

FUTURE INDICATIONS FOR CFH AS THERAPEUTIC COMPLEMENT INHIBITOR

As already outlined, the results of the present work suggested that CFH therapy may be indicated in pathologies requiring CFH's regulatory role for fluid phase regulation of complement.

One etiologically complex pathologic condition shown to involve systemic and local complement-mediated inflammation in its pathogenesis is AMD (Machalinska et al, 2012). The Y402H polymorphism in CFH was discovered to be a major risk allele for AMD (Edwards et al, 2005; Hageman et al, 2005b; Haines et al, 2005; Klein et al, 2005). However, the mechanism how this polymorphism predisposes to AMD remains unclear but is assumed to result in impaired CFH activity as supported by various *in vitro* data (Clark et al, 2010a; Clark et al, 2010b; Weismann et al, 2011).

Therapeutic targeting of complement to intervene early in the course of AMD to inhibit the chronic inflammatory response has been proposed (Troutbeck et al, 2012). Among successful studies on pharmacological complement intervention in AMD models, a recombinant form of human CFH effectively reduced lesion size in a mouse model of wet AMD after systemic administration (Rohrer et al, 2012). Likewise, intravitreal application of purified human plasma CFH in rats undergoing laser-induced choroidal neovascularization (CNV), a model of wet AMD, significantly protected animals from clinical outcomes (Kim et al, 2012).

Thus, these data provide promise in expanding research on the potential of CFH as therapeutic complement inhibitor in human diseases.

9 REFERENCES

Alexander JJ, Quigg RJ (2007) The simple design of complement factor H: Looks can be deceiving. *Mol Immunol* 44: 123-132

Alsenz J, Schulz TF, Lambris JD, Sim RB, Dierich MP (1985) Structural and functional analysis of the complement component factor H with the use of different enzymes and monoclonal antibodies to factor H. *Biochem J* 232: 841-850

Amara U, Flierl MA, Rittirsch D, Klos A, Chen H, Acker B, Bruckner UB, Nilsson B, Gebhard F, Lambris JD, Huber-Lang M (2010) Molecular intercommunication between the complement and coagulation systems. *J Immunol* 185: 5628-5636

Anderson DH, Radeke MJ, Gallo NB, Chapin EA, Johnson PT, Curletti CR, Hancox LS, Hu J, Ebright JN, Malek G, Hauser MA, Rickman CB, Bok D, Hageman GS, Johnson LV (2010) The pivotal role of the complement system in aging and age-related macular degeneration: hypothesis re-visited. *Prog Retin Eye Res* 29: 95-112

Andoh A, Fujiyama Y, Sakumoto H, Uchihara H, Kimura T, Koyama S, Bamba T (1998) Detection of complement C3 and factor B gene expression in normal colorectal mucosa, adenomas and carcinomas. *Clin Exp Immunol* 111: 477-483

Artursson P, Palm K, Luthman K (2012) Caco-2 monolayers in experimental and theoretical predictions of drug transport. *Adv Drug Deliv Rev*

Asgari E, Zhou W, Sacks S (2010) Complement in organ transplantation. *Curr Opin Organ Transplant* 15: 486-491

Aslam M, Perkins SJ (2001) Folded-back solution structure of monomeric factor H of human complement by synchrotron X-ray and neutron scattering, analytical ultracentrifugation and constrained molecular modelling. *J Mol Biol* 309: 1117-1138

Atkinson C, He S, Morris K, Qiao F, Casey S, Goddard M, Tomlinson S (2010) Targeted complement inhibitors protect against posttransplant cardiac ischemia and reperfusion injury and reveal an important role for the alternative pathway of complement activation. *J Immunol* 185: 7007-7013

Atkinson C, Song H, Lu B, Qiao F, Burns TA, Holers VM, Tsokos GC, Tomlinson S (2005a) Targeted complement inhibition by C3d recognition ameliorates tissue injury without apparent increase in susceptibility to infection. *J Clin Invest* 115: 2444-2453

Atkinson C, Song H, Lu B, Qiao F, Burns TA, Holers VM, Tsokos GC, Tomlinson S (2005b) Targeted complement inhibition by C3d recognition ameliorates tissue injury without apparent increase in susceptibility to infection. *J Clin Invest* 115: 2444-2453

Austen WG, Jr., Kobzik L, Carroll MC, Hechtman HB, Moore FD, Jr. (2003) The role of complement and natural antibody in intestinal ischemia-reperfusion injury. *Int J Immunopathol Pharmacol* 16: 1-8

Austen WG, Jr., Zhang M, Chan R, Friend D, Hechtman HB, Carroll MC, Moore FD, Jr. (2004) Murine hindlimb reperfusion injury can be initiated by a self-reactive monoclonal IgM. *Surgery* 136: 401-406

Banda NK, Levitt B, Glogowska MJ, Thurman JM, Takahashi K, Stahl GL, Tomlinson S, Arend WP, Holers VM (2009) Targeted inhibition of the complement alternative pathway with complement receptor 2 and factor H attenuates collagen antibody-induced arthritis in mice. *J Immunol* 183: 5928-5937

Banz Y, Rieben R (2011) Role of complement and perspectives for intervention in ischemia-reperfusion damage. *Ann Med*

Beuk RJ, Heineman E, Tangelder GJ, Kurvers HA, Bonke HJ, oude Egbrink MG (1997) Effects of different durations of total warm ischemia of the gut on rat mesenteric microcirculation. *J Surg Res* 73: 14-23

Beuk RJ, Heineman E, Tangelder GJ, Quaedackers JS, Marks WH, Lieberman JM, oude Egbrink MG (2000) Total warm ischemia and reperfusion impairs flow in all rat gut layers but increases leukocyte-vessel wall interactions in the submucosa only. *Ann Surg* 231: 96-104

Blikslager AT, Moeser AJ, Gookin JL, Jones SL, Odle J (2007) Restoration of barrier function in injured intestinal mucosa. *Physiol Rev* 87: 545-564

Boros M, Takaichi S, Hatanaka K (1995) Ischemic time-dependent microvascular changes and reperfusion injury in the rat small intestine. *J Surg Res* 59: 311-320

Boyuk A, Onder A, Kapan M, Gumus M, Fiotarat U, Basaraliota MK, Alp H (2011) Ellagic acid ameliorates lung injury after intestinal ischemia-reperfusion. *Pharmacogn Mag* 7: 224-228

Bradford MM (1976) A rapid and sensitive method for the quantitation of microgram quantities of protein utilizing the principle of protein-dye binding. *Anal Biochem* 72: 248-254

Brandstatter H, Schulz P, Polunic I, Kannicht C, Kohla G, Romisch J (2012) Purification and biochemical characterization of functional complement factor H from human plasma fractions. *Vox Sang* 103: 201-212

Brooimans RA, van der Ark AA, Buurman WA, van Es LA, Daha MR (1990) Differential regulation of complement factor H and C3 production in human umbilical vein endothelial cells by IFN-gamma and IL-1. *J Immunol* 144: 3835-3840

Busche MN, Stahl GL (2010) Role of the complement components C5 and C3a in a mouse model of myocardial ischemia and reperfusion injury. *Ger Med Sci* 8

Buttner-Mainik A, Parsons J, Jerome H, Hartmann A, Lamer S, Schaaf A, Schlosser A, Zipfel PF, Reski R, Decker EL (2011) Production of biologically active recombinant human factor H in Physcomitrella. *Plant Biotechnol J* 9: 373-383

Carroll MC (2004a) The complement system in regulation of adaptive immunity. *Nat Immunol* 5: 981-986

Carroll MC (2004b) A protective role for innate immunity in systemic lupus erythematosus. *Nat Rev Immunol* 4: 825-831

Chen M, Forrester JV, Xu H (2007) Synthesis of complement factor H by retinal pigment epithelial cells is down-regulated by oxidized photoreceptor outer segments. *Exp Eye Res* 84: 635-645

Chen Y, Lui VC, Rooijen NV, Tam PK (2004) Depletion of intestinal resident macrophages prevents ischaemia reperfusion injury in gut. *Gut* 53: 1772-1780

Chintala MS, Bernardino V, Chiu PJ (1994) Cyclic GMP but not cyclic AMP prevents renal platelet accumulation after ischemia-reperfusion in anesthetized rats. *J Pharmacol Exp Ther* 271: 1203-1208

Chiu CJ, McArdle AH, Brown R, Scott HJ, Gurd FN (1970) Intestinal mucosal lesion in low-flow states. I. A morphological, hemodynamic, and metabolic reappraisal. *Arch Surg* 101: 478-483

Clark SJ, Bishop PN, Day AJ (2010a) Complement factor H and age-related macular degeneration: the role of glycosaminoglycan recognition in disease pathology. *Biochem Soc Trans* 38: 1342-1348

Clark SJ, Perveen R, Hakobyan S, Morgan BP, Sim RB, Bishop PN, Day AJ (2010b) Impaired binding of the age-related macular degeneration-associated complement factor H 402H allotype to Bruch's membrane in human retina. *J Biol Chem* 285: 30192-30202

Coffey PJ, Gias C, McDermott CJ, Lundh P, Pickering MC, Sethi C, Bird A, Fitzke FW, Maass A, Chen LL, Holder GE, Luthert PJ, Salt TE, Moss SE, Greenwood J (2007) Complement factor H deficiency in aged mice causes retinal abnormalities and visual dysfunction. *Proc Natl Acad Sci U S A* 104: 16651-16656

Collard CD, Gelman S (2001) Pathophysiology, clinical manifestations, and prevention of ischemia-reperfusion injury. *Anesthesiology* 94: 1133-1138

Costabile M (2010) Measuring the 50% haemolytic complement (CH50) activity of serum. *J Vis Exp*

Daha MR, Stuffers-Heiman M, Kijlstra A, van Es LA (1979) Isolation and characterization of the third component of rat complement. *Immunology* 36: 63-70

Daha MR, van Es LA (1982) Isolation, characterization, and mechanism of action of rat beta 1H. *J Immunol* 128: 1839-1843

Davin JC, Olie KH, Verlaak R, Horuz F, Florquin S, Weening JJ, Groothoff JW, Strain L, Goodship TH (2006) Complement factor H-associated atypical hemolytic uremic syndrome in monozygotic twins: concordant presentation, discordant response to treatment. *Am J Kidney Dis* 47: e27-30

Deitch EA, Berg R, Specian R (1987) Endotoxin promotes the translocation of bacteria from the gut. *Arch Surg* 122: 185-190

Demberg T, Heine I, Gotze O, Altermann WW, Seliger B, Schlaf G (2006) Recombinant generation of two fragments of the rat complement inhibitory factor H [FH(SCR1-7) and FH(SCR1-4)] and their structural and functional characterization in comparison to FH isolated from rat serum. *Histol Histopathol* 21: 93-102

Demberg T, Pollok-Kopp B, Gerke D, Gotze O, Schlaf G (2002) Rat complement factor H: molecular cloning, sequencing and quantification with a newly established ELISA. *Scand J Immunol* 56: 149-160

DiScipio RG (1992) Ultrastructures and interactions of complement factors H and I. *J Immunol* 149: 2592-2599

Dorweiler B, Pruefer D, Andrasi TB, Maksan SM, Schmiedt W, Neufang A, Vahl CF (2007) Ischemia-Reperfusion Injury. *European Journal of Trauma and Emergency Surgery* 33: 600-612

Edwards AO, Ritter R, 3rd, Abel KJ, Manning A, Panhuysen C, Farrer LA (2005) Complement factor H polymorphism and age-related macular degeneration. *Science* 308: 421-424

Ehrnthaller C, Amara U, Weckbach S, Kalbitz M, Huber-Lang M, Bahrami S (2012) Alteration of complement hemolytic activity in different trauma and sepsis models. *J Inflamm Res* 5: 59-66

Eror AT, Stojadinovic A, Starnes BW, Makrides SC, Tsokos GC, Shea-Donohue T (1999) Antiinflammatory effects of soluble complement receptor type 1 promote rapid recovery of ischemia/reperfusion injury in rat small intestine. *Clin Immunol* 90: 266-275

Fakhouri F, de Jorge EG, Brune F, Azam P, Cook HT, Pickering MC (2010a) Treatment with human complement factor H rapidly reverses renal complement deposition in factor H-deficient mice. *Kidney Int* 78: 279-286

Fakhouri F, de Jorge EG, Brune F, Azam P, Cook HT, Pickering MC (2010b) Treatment with human complement factor H rapidly reverses renal complement deposition in factor H-deficient mice. *Kidney Int* 78: 279-286

Fenaille F, Le Mignon M, Groseil C, Ramon C, Riande S, Siret L, Bihoreau N (2007) Site-specific N-glycan characterization of human complement factor H. *Glycobiology* 17: 932-944

Fleming S (2003) C5a causes limited, polymorphonuclear cell-independent, mesenteric ischemia/reperfusion-induced injury*,**. *Clinical Immunology* 108: 263-273

Fleming SD, Pope MR, Hoffman SM, Moses T, Bukovnik U, Tomich JM, Wagner LM, Woods KM (2010) Domain V peptides inhibit beta2-glycoprotein I-mediated mesenteric ischemia/reperfusion-induced tissue damage and inflammation. *J Immunol* 185: 6168-6178

Fremeaux-Bacchi V, Kemp EJ, Goodship JA, Dragon-Durey MA, Strain L, Loirat C, Deng HW, Goodship TH (2005) The development of atypical haemolytic-uraemic syndrome is influenced by susceptibility factors in factor H and membrane cofactor protein: evidence from two independent cohorts. *J Med Genet* 42: 852-856

Gardner ML, Steele DL (1989) Is there circadian variation in villus height in rat small intestine? *Q J Exp Physiol* 74: 257-265

Gorsuch WB, Chrysanthou E, Schwaeble WJ, Stahl GL (2012) The complement system in ischemia-reperfusion injuries. *Immunobiology* 217: 1026-1033

Granoff DM, Welsch JA, Ram S (2009) Binding of complement factor H (fH) to Neisseria meningitidis is specific for human fH and inhibits complement activation by rat and rabbit sera. *Infect Immun* 77: 764-769

Griffiths MR, Neal JW, Fontaine M, Das T, Gasque P (2009) Complement factor H, a marker of self protects against experimental autoimmune encephalomyelitis. *J Immunol* 182: 4368-4377

Grootjans J, Lenaerts K, Derikx JP, Matthijsen RA, de Bruine AP, van Bijnen AA, van Dam RM, Dejong CH, Buurman WA (2010) Human intestinal ischemia-reperfusion-induced inflammation characterized: experiences from a new translational model. *Am J Pathol* 176: 2283-2291

Grossmann J, Walther K, Artinger M, Rummele P, Woenckhaus M, Scholmerich J (2002) Induction of apoptosis before shedding of human intestinal epithelial cells. *Am J Gastroenterol* 97: 1421-1428

Guan Y, Worrell RT, Pritts TA, Montrose MH (2009) Intestinal ischemia-reperfusion injury: reversible and irreversible damage imaged in vivo. *Am J Physiol Gastrointest Liver Physiol* 297: G187-196

Hageman GS, Anderson DH, Johnson LV, Hancox LS, Taiber AJ, Hardisty LI, Hageman JL, Stockman HA, Borchardt JD, Gehrs KM, Smith RJ, Silvestri G, Russell SR, Klaver CC, Barbazetto I, Chang S, Yannuzzi LA, Barile GR, Merriam JC, Smith RT, Olsh AK, Bergeron J, Zernant J, Merriam JE, Gold B, Dean M, Allikmets R (2005a) A common haplotype in the complement regulatory gene factor H (HF1/CFH) predisposes individuals to age-related macular degeneration. *Proc Natl Acad Sci U S A* 102: 7227-7232

Hageman GS, Anderson DH, Johnson LV, Hancox LS, Taiber AJ, Hardisty LI, Hageman JL, Stockman HA, Borchardt JD, Gehrs KM, Smith RJ, Silvestri G, Russell SR, Klaver CC, Barbazetto I, Chang S, Yannuzzi LA, Barile GR, Merriam JC, Smith RT, Olsh AK, Bergeron J, Zernant J, Merriam JE, Gold B, Dean M, Allikmets R (2005b) A common haplotype in the complement regulatory gene factor H (HF1/CFH) predisposes individuals to age-related macular degeneration. *Proc Natl Acad Sci U S A* 102: 7227-7232

Haines JL, Hauser MA, Schmidt S, Scott WK, Olson LM, Gallins P, Spencer KL, Kwan SY, Noureddine M, Gilbert JR, Schnetz-Boutaud N, Agarwal A, Postel EA, Pericak-Vance MA (2005) Complement factor H variant increases the risk of age-related macular degeneration. *Science* 308: 419-421

Hakobyan S, Harris CL, Tortajada A, Goicochea de Jorge E, Garcia-Layana A, Fernandez-Robredo P, Rodriguez de Cordoba S, Morgan BP (2008) Measurement of factor H variants in plasma using variant-specific monoclonal antibodies: application to assessing risk of age-related macular degeneration. *Invest Ophthalmol Vis Sci* 49: 1983-1990

Hakobyan S, Tortajada A, Harris CL, de Cordoba SR, Morgan BP (2010) Variant-specific quantification of factor H in plasma identifies null alleles associated with atypical hemolytic uremic syndrome. *Kidney Int* 78: 782-788

Hamar J, Racz I, Ciz M, Lojek A, Pallinger E, Furesz J (2003) Time course of leukocyte response and free radical release in an early reperfusion injury of the superior mesenteric artery. *Physiol Res* 52: 417-423

Han WK, Bailly V, Abichandani R, Thadhani R, Bonventre JV (2002) Kidney Injury Molecule-1 (KIM-1): a novel biomarker for human renal proximal tubule injury. *Kidney Int* 62: 237-244

Harboe M, Garred P, Borgen MS, Stahl GL, Roos A, Mollnes TE (2006) Design of a complement mannose-binding lectin pathway-specific activation system applicable at low serum dilutions. *Clin Exp Immunol* 144: 512-520

Harboe M, Mollnes TE (2008) The alternative complement pathway revisited. *J Cell Mol Med* 12: 1074-1084

Harboe M, Ulvund G, Vien L, Fung M, Mollnes TE (2004) The quantitative role of alternative pathway amplification in classical pathway induced terminal complement activation. *Clin Exp Immunol* 138: 439-446

Harhausen D, Khojasteh U, Stahel PF, Morgan BP, Nietfeld W, Dirnagl U, Trendelenburg G (2010) Membrane attack complex inhibitor CD59a protects against focal cerebral ischemia in mice. *J Neuroinflammation* 7: 15

Heinen S, Pluthero FG, van Eimeren VF, Quaggin SE, Licht C (2012) Monitoring and modeling treatment of atypical hemolytic uremic syndrome. *Mol Immunol* 54: 84-88

Hernandez LA, Grisham MB, Twohig B, Arfors KE, Harlan JM, Granger DN (1987) Role of neutrophils in ischemia-reperfusion-induced microvascular injury. *Am J Physiol* 253: H699-703

Hewitt SM, Lewis FA, Cao Y, Conrad RC, Cronin M, Danenberg KD, Goralski TJ, Langmore JP, Raja RG, Williams PM, Palma JF, Warrington JA (2008) Tissue handling and specimen preparation in surgical pathology: issues concerning the recovery of nucleic acids from formalin-fixed, paraffin-embedded tissue. *Arch Pathol Lab Med* 132: 1929-1935

Hill J, Lindsay TF, Ortiz F, Yeh CG, Hechtman HB, Moore FD, Jr. (1992) Soluble complement receptor type 1 ameliorates the local and remote organ injury after intestinal ischemia-reperfusion in the rat. *J Immunol* 149: 1723-1728

Hirt-Minkowski P, Dickenmann M, Schifferli JA (2010) Atypical hemolytic uremic syndrome: update on the complement system and what is new. *Nephron Clin Pract* 114: c219-235

Horton JW, Walker PB (1993) Oxygen radicals, lipid peroxidation, and permeability changes after intestinal ischemia and reperfusion. *J Appl Physiol* 74: 1515-1520

Hourcade DE, Mitchell L, Kuttner-Kondo LA, Atkinson JP, Medof ME (2002) Decay-accelerating factor (DAF), complement receptor 1 (CR1), and factor H dissociate the complement AP C3 convertase (C3bBb) via sites on the type A domain of Bb. *J Biol Chem* 277: 1107-1112

Huang Y, Qiao F, Atkinson C, Holers VM, Tomlinson S (2008a) A novel targeted inhibitor of the alternative pathway of complement and its therapeutic application in ischemia/reperfusion injury. *J Immunol* 181: 8068-8076

Huang Y, Qiao F, Atkinson C, Holers VM, Tomlinson S (2008b) A novel targeted inhibitor of the alternative pathway of complement and its therapeutic application in ischemia/reperfusion injury. *J Immunol* 181: 8068-8076

Huber-Lang M, Sarma JV, Zetoune FS, Rittirsch D, Neff TA, McGuire SR, Lambris JD, Warner RL, Flierl MA, Hoesel LM, Gebhard F, Younger JG, Drouin SM, Wetsel RA, Ward PA (2006) Generation of C5a in the absence of C3: a new complement activation pathway. *Nat Med* 12: 682-687

Iizuka M, Konno S (2011) Wound healing of intestinal epithelial cells. *World J Gastroenterol* 17: 2161-2171

Ikeda H, Suzuki Y, Suzuki M, Koike M, Tamura J, Tong J, Nomura M, Itoh G (1998) Apoptosis is a major mode of cell death caused by ischaemia and ischaemia/reperfusion injury to the rat intestinal epithelium. *Gut* 42: 530-537

Jokiranta TS, Jaakola VP, Lehtinen MJ, Parepalo M, Meri S, Goldman A (2006) Structure of complement factor H carboxyl-terminus reveals molecular basis of atypical haemolytic uremic syndrome. *EMBO J* 25: 1784-1794

Jouvin MH, Kazatchkine MD, Cahour A, Bernard N (1984) Lysine residues, but not carbohydrates, are required for the regulatory function of H on the amplification C3 convertase of complement. *Journal of immunology* 133: 3250-3254

Kalia N, Pockley AG, Wood RF, Brown NJ (2002) Effects of hypothermia and rewarming on the mucosal villus microcirculation and survival after rat intestinal ischemia-reperfusion injury. *Ann Surg* 236: 67-74

Katz Y, Strunk RC (1988) Synthesis and regulation of complement protein factor H in human skin fibroblasts. *J Immunol* 141: 559-563

Kim SJ, Kim J, Lee J, Cho SY, Kang HJ, Kim KY, Jin DK (2012) Intravitreal human complement factor H in a rat model of laser-induced choroidal neovascularisation. *Br J Ophthalmol*

Kim YH, He S, Kase S, Kitamura M, Ryan SJ, Hinton DR (2009) Regulated secretion of complement factor H by RPE and its role in RPE migration. *Graefes Arch Clin Exp Ophthalmol* 247: 651-659

Klein RJ, Zeiss C, Chew EY, Tsai JY, Sackler RS, Haynes C, Henning AK, SanGiovanni JP, Mane SM, Mayne ST, Bracken MB, Ferris FL, Ott J, Barnstable C, Hoh J (2005) Complement factor H polymorphism in age-related macular degeneration. *Science* 308: 385-389

Koistinen V (1991) Effect of complement-protein-C3b density on the binding of complement factor H to surface-bound C3b. *Biochem J* 280 (Pt 1): 255-259

Kulik L, Fleming SD, Moratz C, Reuter JW, Novikov A, Chen K, Andrews KA, Markaryan A, Quigg RJ, Silverman GJ, Tsokos GC, Holers VM (2009) Pathogenic natural antibodies recognizing annexin IV are required to develop intestinal ischemia-reperfusion injury. *J Immunol* 182: 5363-5373

Lambris JD, Lao Z, Oglesby TJ, Atkinson JP, Hack CE, Becherer JD (1996) Dissection of CR1, factor H, membrane cofactor protein, and factor B binding and functional sites in the third complement component. *Journal of immunology* 156: 4821-4832

Lambris JD, Ricklin D, Geisbrecht BV (2008) Complement evasion by human pathogens. *Nat Rev Microbiol* 6: 132-142

Lauer N, Mihlan M, Hartmann A, Schlotzer-Schrehardt U, Keilhauer C, Scholl HP, Charbel Issa P, Holz F, Weber BH, Skerka C, Zipfel PF (2011) Complement regulation at necrotic cell lesions is impaired by the age-related macular degeneration-associated factor-H His402 risk variant. *J Immunol* 187: 4374-4383

Lee H, Green DJ, Lai L, Hou YJ, Jensenius JC, Liu D, Cheong C, Park CG, Zhang M (2010) Early complement factors in the local tissue immunocomplex generated during intestinal ischemia/reperfusion injury. *Mol Immunol* 47: 972-981

Leffler J, Herbert AP, Norstrom E, Schmidt CQ, Barlow PN, Blom AM, Martin M (2010) Annexin-II, DNA, and histones serve as factor H ligands on the surface of apoptotic cells. *J Biol Chem* 285: 3766-3776

Leinhase I, Schmidt OI, Thurman JM, Hossini AM, Rozanski M, Taha ME, Scheffler A, John T, Smith WR, Holers VM, Stahel PF (2006) Pharmacological complement inhibition at the C3 convertase level promotes neuronal survival, neuroprotective intracerebral gene expression, and neurological outcome after traumatic brain injury. *Exp Neurol* 199: 454-464

Leite Junior R, Mello NB, Pereira Lde P, Takiya CM, Oliveira CA, Schanaider A (2010) Enterocyte ultrastructural alterations following intestinal obstruction in rats. *Acta Cir Bras* 25: 2-8

Licht C, Schlotzer-Schrehardt U, Kirschfink M, Zipfel PF, Hoppe B (2007) MPGN II--genetically determined by defective complement regulation? *Pediatr Nephrol* 22: 2-9

Lu F, Chauhan AK, Fernandes SM, Walsh MT, Wagner DD, Davis AE, 3rd (2008) The effect of C1 inhibitor on intestinal ischemia and reperfusion injury. *Am J Physiol Gastrointest Liver Physiol* 295: G1042-1049

Lu YZ, Wu CC, Huang YC, Huang CY, Yang CY, Lee TC, Chen CF, Yu LC (2012) Neutrophil priming by hypoxic preconditioning protects against epithelial barrier damage and enteric bacterial translocation in intestinal ischemia/reperfusion. *Lab Invest* 92: 783-796

Lucas JG, Co JP, Nwaogwugwu UT, Dosani I, Sureshkumar KK (2011) Antibody-mediated rejection in kidney transplantation: an update. *Expert Opin Pharmacother* 12: 579-592

Lundh von Leithner P, Kam JH, Bainbridge J, Catchpole I, Gough G, Coffey P, Jeffery G (2009) Complement factor h is critical in the maintenance of retinal perfusion. *Am J Pathol* 175: 412-421

Machalinska A, Kawa MP, Marlicz W, Machalinski B (2012) Complement system activation and endothelial dysfunction in patients with age-related macular degeneration (AMD): possible relationship between AMD and atherosclerosis. *Acta Ophthalmol* 90: 695-703

Mallick IH, Yang W, Winslet MC, Seifalian AM (2004) Ischemia-reperfusion injury of the intestine and protective strategies against injury. *Dig Dis Sci* 49: 1359-1377

Mandal MN, Ayyagari R (2006) Complement factor H: spatial and temporal expression and localization in the eye. *Invest Ophthalmol Vis Sci* 47: 4091-4097

Mangus R, Vianna R, Tector A (2009) Intestinal transplantation: an overview. *Minerva Chir* 64: 45-57

Martin M, Leffler J, Blom AM (2012) Annexin A2 and a5 serve as new ligands for c1q on apoptotic cells. *J Biol Chem* 287: 33733-33744

Massberg S, Gonzalez AP, Leiderer R, Menger MD, Messmer K (1998) In vivo assessment of the influence of cold preservation time on microvascular reperfusion injury after experimental small bowel transplantation. *Br J Surg* 85: 127-133

Matthijsen RA, Derikx JP, Kuipers D, van Dam RM, Dejong CH, Buurman WA (2009) Enterocyte shedding and epithelial lining repair following ischemia of the human small intestine attenuate inflammation. *PLoS One* 4: e7045

Mayer MM (1961) On the destruction of erythrocytes and other cells by antibody and complement. *Cancer Res* 21: 1262-1269

Mayilyan KR (2012) Complement genetics, deficiencies, and disease associations. *Protein Cell* 3: 487-496

McRae JL, Duthy TG, Griggs KM, Ormsby RJ, Cowan PJ, Cromer BA, McKinstry WJ, Parker MW, Murphy BF, Gordon DL (2005) Human factor H-related protein 5 has cofactor activity, inhibits C3 convertase activity, binds heparin and C-reactive protein, and associates with lipoprotein. *J Immunol* 174: 6250-6256

Meng FW, Shimoda H, Kajiwara T, Matsuda M, Kato S (2007) Reconstruction of central lacteals in the murine jejunum following ischemia-reperfusion injury. *Arch Histol Cytol* 70: 135-146

Meri S, Pangburn MK (1990) Discrimination between activators and nonactivators of the alternative pathway of complement: regulation via a sialic acid/polyanion binding site on factor H. *Proc Natl Acad Sci U S A* 87: 3982-3986

Mihlan M, Stippa S, Jozsi M, Zipfel PF (2009) Monomeric CRP contributes to complement control in fluid phase and on cellular surfaces and increases phagocytosis by recruiting factor H. *Cell Death Differ* 16: 1630-1640

Moon R, Parikh AA, Szabo C, Fischer JE, Salzman AL, Hasselgren PO (1997) Complement C3 production in human intestinal epithelial cells is regulated by interleukin 1beta and tumor necrosis factor alpha. *Arch Surg* 132: 1289-1293

Muller-Eberhard HJ (1986) The membrane attack complex of complement. *Annu Rev Immunol* 4: 503-528

Nakamoto M, Shapiro JI, Shanley PF, Chan L, Schrier RW (1987) In vitro and in vivo protective effect of atriopeptin III on ischemic acute renal failure. *J Clin Invest* 80: 698-705

Nauta AJ, Trouw LA, Daha MR, Tijsma O, Nieuwland R, Schwaeble WJ, Gingras AR, Mantovani A, Hack EC, Roos A (2002) Direct binding of C1q to apoptotic cells and cell blebs induces complement activation. *Eur J Immunol* 32: 1726-1736

Nilsson B, Nilsson Ekdahl K (2012) The tick-over theory revisited: is C3 a contact-activated protein? *Immunobiology* 217: 1106-1110

Oikonomopoulou K, Ricklin D, Ward PA, Lambris JD (2012) Interactions between coagulation and complement--their role in inflammation. *Semin Immunopathol* 34: 151-165

Okemefuna AI, Nan R, Gor J, Perkins SJ (2009) Electrostatic interactions contribute to the folded-back conformation of wild type human factor H. *J Mol Biol* 391: 98-118

Padilla ND, van Vliet AK, Schoots IG, Valls Seron M, Maas MA, Peltenburg EE, de Vries A, Niessen HW, Hack CE, van Gulik TM (2007) C-reactive protein and natural IgM antibodies are activators of complement in a rat model of intestinal ischemia and reperfusion. *Surgery* 142: 722-733

Paixao-Cavalcante D, Hanson S, Botto M, Cook HT, Pickering MC (2009) Factor H facilitates the clearance of GBM bound iC3b by controlling C3 activation in fluid phase. *Molecular immunology* 46: 1942-1950

Palinski W, Yla-Herttuala S, Rosenfeld ME, Butler SW, Socher SA, Parthasarathy S, Curtiss LK, Witztum JL (1990) Antisera and monoclonal antibodies specific for epitopes generated during oxidative modification of low density lipoprotein. *Arteriosclerosis* 10: 325-335

Pangburn MK, Rawal N, Cortes C, Alam MN, Ferreira VP, Atkinson MA (2009) Polyanion-induced self-association of complement factor H. *J Immunol* 182: 1061-1068

Pechtl IC, Kavanagh D, McIntosh N, Harris CL, Barlow PN (2011) Disease-associated N-terminal complement factor H mutations perturb cofactor and decay-accelerating activities. *J Biol Chem* 286: 11082-11090

Pemberton M, Anderson G, Vetvicka V, Justus DE, Ross GD (1993) Microvascular effects of complement blockade with soluble recombinant CR1 on ischemia/reperfusion injury of skeletal muscle. *J Immunol* 150: 5104-5113

Perkins SJ, Nan R, Okemefuna AI, Li K, Khan S, Miller A (2010) Multiple interactions of complement Factor H with its ligands in solution: a progress report. *Adv Exp Med Biol* 703: 25-47

Perkins SJ, Nealis AS, Sim RB (1991) Oligomeric domain structure of human complement factor H by X-ray and neutron solution scattering. *Biochemistry* 30: 2847-2857

Pickering MC, Warren J, Rose KL, Carlucci F, Wang Y, Walport MJ, Cook HT, Botto M (2006) Prevention of C5 activation ameliorates spontaneous and experimental glomerulonephritis in factor H-deficient mice. *Proc Natl Acad Sci U S A* 103: 9649-9654

Pope MR, Bukovnik U, Tomich JM, Fleming SD (2012) Small beta2-glycoprotein I peptides protect from intestinal ischemia reperfusion injury. *J Immunol* 189: 5047-5056

Rehrig S, Fleming SD, Anderson J, Guthridge JM, Rakstang J, McQueen CE, Holers VM, Tsokos GC, Shea-Donohue T (2001) Complement inhibitor, complement receptor 1-related gene/protein y-Ig attenuates intestinal damage after the onset of mesenteric ischemia/reperfusion injury in mice. *J Immunol* 167: 5921-5927

Renner B, Coleman K, Goldberg R, Amura C, Holland-Neidermyer A, Pierce K, Orth HN, Molina H, Ferreira VP, Cortes C, Pangburn MK, Holers VM, Thurman JM (2010) The complement inhibitors Crry and factor H are critical for preventing autologous complement activation on renal tubular epithelial cells. *J Immunol* 185: 3086-3094

Renner B, Ferreira VP, Cortes C, Goldberg R, Ljubanovic D, Pangburn MK, Pickering MC, Tomlinson S, Holland-Neidermyer A, Strassheim D, Holers VM, Thurman JM (2011) Binding of factor H to tubular epithelial cells limits interstitial complement activation in ischemic injury. *Kidney Int* 80: 165-173

Ricklin D, Hajishengallis G, Yang K, Lambris JD (2010) Complement: a key system for immune surveillance and homeostasis. *Nat Immunol* 11: 785-797

Ripoche J, Erdei A, Gilbert D, Al Salihi A, Sim RB, Fontaine M (1988) Two populations of complement factor H differ in their ability to bind to cell surfaces. *Biochem J* 253: 475-480

Risitano AM, Notaro R, Pascariello C, Sica M, del Vecchio L, Horvath CJ, Fridkis-Hareli M, Selleri C, Lindorfer MA, Taylor RP, Luzzatto L, Holers VM (2012) The complement receptor 2/factor H fusion protein TT30 protects paroxysmal nocturnal hemoglobinuria erythrocytes from complement-mediated hemolysis and C3 fragment. *Blood* 119: 6307-6316

Rodrigues SF, Granger DN (2010) Role of blood cells in ischaemia-reperfusion induced endothelial barrier failure. *Cardiovasc Res* 87: 291-299

Rohrer B, Coughlin B, Bandyopadhyay M, Holers VM (2012) Systemic human CR2-targeted complement alternative pathway inhibitor ameliorates mouse laser-induced choroidal neovascularization. *J Ocul Pharmacol Ther* 28: 402-409

Ross GD, Newman SL, Lambris JD, Devery-Pocius JE, Cain JA, Lachmann PJ (1983) Generation of three different fragments of bound C3 with purified factor I or serum. II. Location of binding sites in the C3 fragments for factors B and H, complement receptors, and bovine conglutinin. *J Exp Med* 158: 334-352

Saito A (2008) Plasma kallikrein is activated on dermatan sulfate and cleaves factor H. *Biochem Biophys Res Commun* 370: 646-650

Saito A, Munakata H (2005) Factor H is a dermatan sulfate-binding protein: identification of a dermatan sulfate-mediated protease that cleaves factor H. *J Biochem* 137: 225-233

Schmidt CQ, Slingsby FC, Richards A, Barlow PN (2011) Production of biologically active complement factor H in therapeutically useful quantities. *Protein Expr Purif* 76: 254-263

Schneider MC, Prosser BE, Caesar JJ, Kugelberg E, Li S, Zhang Q, Quoraishi S, Lovett JE, Deane JE, Sim RB, Roversi P, Johnson S, Tang CM, Lea SM (2009) Neisseria meningitidis recruits factor H using protein mimicry of host carbohydrates. *Nature* 458: 890-893

Sener G, Akgun U, Satiroglu H, Topaloglu U, Keyer-Uysal M (2001) The effect of pentoxifylline on intestinal ischemia/reperfusion injury. *Fundam Clin Pharmacol* 15: 19-22

Serrano J, Encinas JM, Fernandez AP, Castro-Blanco S, Alonso D, Fernandez-Vizarra P, Richart A, Bentura ML, Santacana M, Cuttitta F, Martinez A, Rodrigo J (2003) Distribution of immunoreactivity for the adrenomedullin binding protein, complement factor H, in the rat brain. *Neuroscience* 116: 947-962

Shackelford C, Long G, Wolf J, Okerberg C, Herbert R (2002) Qualitative and quantitative analysis of nonneoplastic lesions in toxicology studies. *Toxicol Pathol* 30: 93-96

Shevchenko A, Wilm M, Vorm O, Mann M (1996) Mass spectrometric sequencing of proteins silver-stained polyacrylamide gels. *Anal Chem* 68: 850-858

Sim RB, DiScipio RG (1982a) Purification and structural studies on the complement-system control protein beta 1H (Factor H). *Biochem J* 205: 285-293

Sim RB, DiScipio RG (1982b) Purification and structural studies on the complement-system control protein beta 1H (Factor H). *Biochem J* 205: 285-293

Simpson R, Alon R, Kobzik L, Valeri CR, Shepro D, Hechtman HB (1993) Neutrophil and nonneutrophil-mediated injury in intestinal ischemia-reperfusion. *Ann Surg* 218: 444-453; discussion 453-444

Smith RJ, Alexander J, Barlow PN, Botto M, Cassavant TL, Cook HT, de Cordoba SR, Hageman GS, Jokiranta TS, Kimberling WJ, Lambris JD, Lanning LD, Levidiotis V, Licht C, Lutz HU, Meri S, Pickering MC, Quigg RJ, Rops AL, Salant DJ, Sethi S, Thurman JM, Tully HF, Tully SP, van der Vlag J, Walker PD, Wurzner R, Zipfel PF (2007) New approaches to the treatment of dense deposit disease. *J Am Soc Nephrol* 18: 2447-2456

Souza DG, Esser D, Bradford R, Vieira AT, Teixeira MM (2005) APT070 (Mirococept), a membrane-localised complement inhibitor, inhibits inflammatory responses that follow intestinal ischaemia and reperfusion injury. *Br J Pharmacol* 145: 1027-1034

Sparkes BL, Slone EE, Roth M, Welti R, Fleming SD (2010) Intestinal lipid alterations occur prior to antibody-induced prostaglandin E2 production in a mouse model of ischemia/reperfusion. *Biochimica et biophysica acta* 1801: 517-525

Stahl GL, Xu Y, Hao L, Miller M, Buras JA, Fung M, Zhao H (2003) Role for the alternative complement pathway in ischemia/reperfusion injury. *Am J Pathol* 162: 449-455

Stallion A, Kou TD, Latifi SQ, Miller KA, Dahms BB, Dudgeon DL, Levine AD (2005) Ischemia/reperfusion: a clinically relevant model of intestinal injury yielding systemic inflammation. *J Pediatr Surg* 40: 470-477

Strohalm M, Kavan D, Novak P, Volny M, Havlicek V (2010) mMass 3: a cross-platform software environment for precise analysis of mass spectrometric data. *Anal Chem* 82: 4648-4651

Szabo A, Vollmar B, Boros M, Menger MD (2008) In vivo fluorescence microscopic imaging for dynamic quantitative assessment of intestinal mucosa permeability in mice. *J Surg Res* 145: 179-185

Tekin IO, Sipahi EY, Comert M, Acikgoz S, Yurdakan G (2009) Low-density lipoproteins oxidized after intestinal ischemia/reperfusion in rats. *J Surg Res* 157: e47-54

Thurman JM, Ljubanovic D, Edelstein CL, Gilkeson GS, Holers VM (2003a) Lack of a functional alternative complement pathway ameliorates ischemic acute renal failure in mice. *J Immunol* 170: 1517-1523

Thurman JM, Ljubanovic D, Edelstein CL, Gilkeson GS, Holers VM (2003b) Lack of a functional alternative complement pathway ameliorates ischemic acute renal failure in mice. *J Immunol* 170: 1517-1523

Thurman JM, Renner B, Kunchithapautham K, Ferreira VP, Pangburn MK, Ablonczy Z, Tomlinson S, Holers VM, Rohrer B (2009) Oxidative stress renders retinal pigment epithelial cells susceptible to complement-mediated injury. *J Biol Chem* 284: 16939-16947

Tichaczek-Goska D (2012) Deficiencies and excessive human complement system activation in disorders of multifarious etiology. *Adv Clin Exp Med* 21: 105-114

Timar KK, Pasch MC, van den Bosch NH, Jarva H, Junnikkala S, Meri S, Bos JD, Asghar SS (2006) Human keratinocytes produce the complement inhibitor factor H: synthesis is regulated by interferon-gamma. *Mol Immunol* 43: 317-325

Tripatara P, Patel NS, Webb A, Rathod K, Lecomte FM, Mazzon E, Cuzzocrea S, Yaqoob MM, Ahluwalia A, Thiemermann C (2007) Nitrite-derived nitric oxide protects the rat kidney against ischemia/reperfusion injury in vivo: role for xanthine oxidoreductase. *J Am Soc Nephrol* 18: 570-580

Troutbeck R, Al-Qureshi S, Guymer RH (2012) Therapeutic targeting of the complement system in age-related macular degeneration: a review. *Clin Experiment Ophthalmol* 40: 18-26

Trouw LA, Bengtsson AA, Gelderman KA, Dahlback B, Sturfelt G, Blom AM (2007) C4b-binding protein and factor H compensate for the loss of membrane-bound complement inhibitors to protect apoptotic cells against excessive complement attack. *J Biol Chem* 282: 28540-28548

Tsunooka N (2004) Bacterial translocation secondary to small intestinal mucosal ischemia during cardiopulmonary bypass. Measurement by diamine oxidase and peptidoglycan. *European Journal of Cardio-Thoracic Surgery* 25: 275-280

Turner JR (2009) Intestinal mucosal barrier function in health and disease. *Nat Rev Immunol* 9: 799-809

Vakeva A, Meri S (1998) Complement activation and regulator expression after anoxic injury of human endothelial cells. *APMIS* 106: 1149-1156

Vollmar B, Menger MD (2011) Intestinal ischemia/reperfusion: microcirculatory pathology and functional consequences. *Langenbecks Arch Surg* 396: 13-29

Wada K, Montalto MC, Stahl GL (2001) Inhibition of complement C5 reduces local and remote organ injury after intestinal ischemia/reperfusion in the rat. *Gastroenterology* 120: 126-133

Weismann D, Hartvigsen K, Lauer N, Bennett KL, Scholl HP, Charbel Issa P, Cano M, Brandstatter H, Tsimikas S, Skerka C, Superti-Furga G, Handa JT, Zipfel PF, Witztum JL, Binder CJ (2011) Complement factor H binds malondialdehyde epitopes and protects from oxidative stress. *Nature* 478: 76-81

Werner M, Chott A, Fabiano A, Battifora H (2000) Effect of formalin tissue fixation and processing on immunohistochemistry. *Am J Surg Pathol* 24: 1016-1019

Whaley K (1980) Biosynthesis of the complement components and the regulatory proteins of the alternative complement pathway by human peripheral blood monocytes. *J Exp Med* 151: 501-516

Whaley K, Ruddy S (1976) Modulation of the alternative complement pathways by beta 1 H globulin. *J Exp Med* 144: 1147-1163

Wilczek E, Rzepko R, Nowis D, Legat M, Golab J, Glab M, Gorlewicz A, Konopacki F, Mazurkiewicz M, Sladowski D, Gornicka B, Wasiutynski A, Wilczynski GM (2008) The possible role of factor H in colon cancer resistance to complement attack. *Int J Cancer* 122: 2030-2037

Williams JP, Pechet TT, Weiser MR, Reid R, Kobzik L, Moore FD, Jr., Carroll MC, Hechtman HB (1999) Intestinal reperfusion injury is mediated by IgM and complement. *J Appl Physiol* 86: 938-942

Yamamoto M, Plessow B, Koch HK, Oehlert W (1980) Electron microscopic studies on the small intestinal mucosa of rats after mechanical intestinal obstruction and ischemia. *Virchows Arch B Cell Pathol Incl Mol Pathol* 32: 157-164

Zhang M, Austen WG, Jr., Chiu I, Alicot EM, Hung R, Ma M, Verna N, Xu M, Hechtman HB, Moore FD, Jr., Carroll MC (2004) Identification of a specific self-reactive IgM antibody that initiates intestinal ischemia/reperfusion injury. *Proc Natl Acad Sci U S A* 101: 3886-3891

Zhang M, Michael LH, Grosjean SA, Kelly RA, Carroll MC, Entman ML (2006) The role of natural IgM in myocardial ischemia-reperfusion injury. *J Mol Cell Cardiol* 41: 62-67

Zipfel PF, Edey M, Heinen S, Jozsi M, Richter H, Misselwitz J, Hoppe B, Routledge D, Strain L, Hughes AE, Goodship JA, Licht C, Goodship TH, Skerka C (2007) Deletion of complement factor H-related genes CFHR1 and CFHR3 is associated with atypical hemolytic uremic syndrome. *PLoS Genet* 3: e41

Zipfel PF, Skerka C (2009) Complement regulators and inhibitory proteins. *Nat Rev Immunol* 9: 729-740

10 LIST OF ABBREVIATIONS

AEX	Anion Exchange Chromatography
AGES	Austrian Agency for Health and Food Safety
aHUS	atypical Hemolytic Uremic Syndrome
ANOVA	Analysis of variance
AP	Alternative Complement Pathway
BUN	Blood Urea Nitrogen
CEX	Cation Exchange Chromatography
CFH	Complement Factor H
CHA	Ceramic Hydroxyapatite Chromatography
CH50	Hemolytic Complement Activity
C3	Complement C3 protein
DAB	diaminobenzidine
EDTA	Ethylene diamine tetraacetic acid
EGTA	Ethylene glycol tetraacetic acid
ELISA	Enzyme-linked Immunosorbent Assay
FBS	Formalin-buffered saline
FFPE	Formalin fixed paraffin embedded
FITC	fluoresceinisothiocyanate
H+E	Hematoxylin & Eosin
HPLC	High Performance Liquid Chromatography
HRP	Horseradish Peroxidase
IgG	Immunoglobulin G

IgM	Immunoglobulin M
IHC	Immunohistochemistry
IRI	Ischemia Reperfusion Injury
i.v.	intravenous
MALDI-TOF	Matrix-ass. laser desorption ionization Time-of-Flight
MBL	Mannose Binding Lectin
MDA	Malondialdehyde
KIM-1	Kidney Injury Molecule-1
MPGN-II	Membranoproliferative Glomerulonephritis type II
NaCl	sodium chloride
NAG	N-acetyl-β-D-glucosaminidase
OCT	Optimum Cutting Temperature compound
oxLDL	Oxidized Low Density Lipoproteins
PBS	Phosphate buffered saline
SCR	Short Consensus Repeat
S/D	Solvent Detergent
SDS-PAGE	Sodium dodecylsulfate – polyacrylamide electrophoresis
SDS	Sodium dodecylsulphate
SEM	Standard Error of the mean
SRBC	Sheep Red Blood Cells
TBS	TRIS-buffered saline
TRIS	tris (hydroxyl-methyl) amino-methane
UDF	Ultradiafiltration

11 PUBLICATIONS

Brandstaetter H. *et al.* Vox Sanguinis 2012 Oct;103(3):201-12. doi: 10.1111/j.1423-0410.2012.01610.x. Epub 2012 Apr 12.

Purification and biochemical characterization of functional complement factor H from human plasma fractions

Brandstaetter H, Schulz P, Roemisch J, International Patent Application, WO 2012/049245 A1; Method for Purification of Complement Factor H

Weismann D. et al. Nature 2011 Oct 5;478(7367):76-81. doi: 10.1038/nature10449.

Complement factor H binds malondialdehyde epitopes and protects from oxidative stress.

Heger A, Brandstätter H, et al. Transfus Apher Sci. 2013 May 22. doi: 10.1016/j.transci.2013.04.039.

Universal pooled plasma (Uniplas®) does not induce complement-mediated hemolysis of human red blood cells *in vitro*

12 LIST OF TABLES

Table 1 – Process parameters of CFH purification process

Table 2 – Purified CFH contains traces of accompanying plasma proteins

Table 3 – No changes of inflammatory cell in rats injected with human CFH

13 LIST OF FIGURES

Figure 1 – The Complement System

Figure 2 – CFH functions within the complement cascade

Figure 3 - CFH preparation process

Figure 4 - Positioning of vascular clamp and sutures

Figure 5 - Grading of Intestinal Injury on behalf of the mucosal appearance (Chiu Score)

Figure 6 - Intestinal IRI in rats on a time scale.

Figure 7 - Surgery to create intestinal IRI in rats.

Figure 8 – Intestinal IRI but not sham surgery induces increased mucosal pathology and decreased villi heights in ileum

Figure 9 - Morphological appearance of rat intestines after sham surgery and intestinal IRI

Figure 10 – Reduced plasma protein, C3 and CH50 in rats

Figure 11 – Local C3 deposition at rat ileum subjected to IRI.

Figure 12 – Human CFH injection does not rescue rats from intestinal IRI

Figure 13 – Detection of human CFH in rat plasma by ELISA

Figure 14 – Specificity of anti-CFH antibodies towards rat and human

Figure 15 – Anti rat CFH antibody detects endogenous CFH in rat intestine, liver and brain

Figure 16 – Endogenous CFH along epithelia in rat ileum

Figure 17 – Human CFH remained in intestinal blood vessels and did not target the epithelium in gut IRI

Figure 18 – Human CFH did not target renal sites of injury after renal IRI

Figure 19 - Caco-2 cell model of human intestinal epithelial cells expresses CFH

Figure 20 – Rat and Human CFH focally bound dying intestinal epithelial cells in intestinal IRI

Figure 21 – Human CFH maintained hemoprotective activity after 90 min within plasma of IRI rats

Figure 22 – Human CFH protected from local complement deposition on ischemic rat ileum

i want morebooks!

Buy your books fast and straightforward online - at one of world's fastest growing online book stores! Environmentally sound due to Print-on-Demand technologies.

Buy your books online at
www.get-morebooks.com

Kaufen Sie Ihre Bücher schnell und unkompliziert online – auf einer der am schnellsten wachsenden Buchhandelsplattformen weltweit! Dank Print-On-Demand umwelt- und ressourcenschonend produziert.

Bücher schneller online kaufen
www.morebooks.de

VDM Verlagsservicegesellschaft mbH
Heinrich-Böcking-Str. 6-8
D - 66121 Saarbrücken
Telefon: +49 681 3720 174
Telefax: +49 681 3720 1749
info@vdm-vsg.de
www.vdm-vsg.de

Printed by Books on Demand GmbH, Norderstedt / Germany